Metropolitan Preoccupations

RGS-IBG Book Series

For further information about the series and a full list of published and forthcoming titles please visit www.rgsbookseries.com

Published

Metropolitan Preoccupations: The Spatial Politics of Squatting in Berlin
Alexander Vasudevan

Everyday Peace? Politics, Citizenship and Muslim Lives in India
Philippa Williams

Assembling Export Markets: The Making and Unmaking of Global Food Connections in West Africa
Stefan Ouma

Africa's Information Revolution: Technical Regimes and Production Networks in South Africa and Tanzania
James T. Murphy and Pádraig Carmody

Origination: The Geographies of Brands and Branding
Andy Pike

In the Nature of Landscape: Cultural Geography on the Norfolk Broads
David Matless

Geopolitics and Expertise: Knowledge and Authority in European Diplomacy
Merje Kuus

Everyday Moral Economies: Food, Politics and Scale in Cuba
Marisa Wilson

Material Politics: Disputes Along the Pipeline
Andrew Barry

Fashioning Globalisation: New Zealand Design, Working Women and the Cultural Economy
Maureen Molloy and Wendy Larner

Working Lives - Gender, Migration and Employment in Britain, 1945–2007
Linda McDowell

Dunes: Dynamics, Morphology and Geological History
Andrew Warren

Spatial Politics: Essays for Doreen Massey
Edited by David Featherstone and Joe Painter

The Improvised State: Sovereignty, Performance and Agency in Dayton Bosnia
Alex Jeffrey

Learning the City: Knowledge and Translocal Assemblage
Colin McFarlane

Globalizing Responsibility: The Political Rationalities of Ethical Consumption
Clive Barnett, Paul Cloke, Nick Clarke and Alice Malpass

Domesticating Neo-Liberalism: Spaces of Economic Practice and Social Reproduction in Post-Socialist Cities
Alison Stenning, Adrian Smith, Alena Rochovská and Dariusz Świątek

Swept Up Lives? Re-envisioning the Homeless City
Paul Cloke, Jon May and Sarah Johnsen

Aerial Life: Spaces, Mobilities, Affects
Peter Adey

Millionaire Migrants: Trans-Pacific Life Lines
David Ley

State, Science and the Skies: Governmentalities of the British Atmosphere
Mark Whitehead

Complex Locations: Women's Geographical Work in the UK 1850–1970
Avril Maddrell

Value Chain Struggles: Institutions and Governance in the Plantation Districts of South India
Jeff Neilson and Bill Pritchard

Queer Visibilities: Space, Identity and Interaction in Cape Town
Andrew Tucker

Arsenic Pollution: A Global Synthesis
Peter Ravenscroft, Hugh Brammer and Keith Richards

Resistance, Space and Political Identities: The Making of Counter-Global Networks
David Featherstone

Mental Health and Social Space: Towards Inclusionary Geographies?
Hester Parr

Climate and Society in Colonial Mexico: A Study in Vulnerability
Georgina H. Endfield

Geochemical Sediments and Landscapes
Edited by David J. Nash and Sue J. McLaren

Driving Spaces: A Cultural-Historical Geography of England's M1 Motorway
Peter Merriman

Badlands of the Republic: Space, Politics and Urban Policy
Mustafa Dikeç

Geomorphology of Upland Peat: Erosion, Form and Landscape Change
Martin Evans and Jeff Warburton

Spaces of Colonialism: Delhi's Urban Governmentalities
Stephen Legg

People/States/Territories
Rhys Jones

Publics and the City
Kurt Iveson

After the Three Italies: Wealth, Inequality and Industrial Change
Mick Dunford and Lidia Greco

Putting Workfare in Place
Peter Sunley, Ron Martin and Corinne Nativel

Domicile and Diaspora
Alison Blunt

Geographies and Moralities
Edited by Roger Lee and David M. Smith

Military Geographies
Rachel Woodward

A New Deal for Transport?
Edited by Iain Docherty and Jon Shaw

Geographies of British Modernity
Edited by David Gilbert, David Matless and Brian Short

Lost Geographies of Power
John Allen

Globalizing South China
Carolyn L. Cartier

Geomorphological Processes and Landscape Change: Britain in the Last 1000 Years
Edited by David L. Higgitt and E. Mark Lee

Forthcoming

Smoking Geographies: Space, Place and Tobacco
Ross Barnett, Graham Moon, Jamie Pearce, Lee Thompson and Liz Twigg

Home SOS: Gender, Injustice and Rights in Cambodia
Katherine Brickell

Pathological Lives: Disease, Space and Biopolitics
Steve Hinchliffe, Nick Bingham, John Allen and Simon Carter

Work–Life Advantage: Sustaining Regional Learning and Innovation
Al James

Rehearsing the State: The Political Practices of the Tibetan Government-in-Exile
Fiona McConnell

Articulations of Capital: Global Production Networks and Regional Transformations
John Pickles, Adrian Smith and Robert Begg, with Milan Buček, Rudolf Pástor and Poli Roukova

Body, Space and Affect
Steve Pile

Making Other Worlds: Agency and Interaction in Environmental Change
John Wainwright

Transnational Geographies of the Heart: Intimacy in a Globalising World
Katie Walsh

Metropolitan Preoccupations

The Spatial Politics of Squatting in Berlin

Alexander Vasudevan

WILEY Blackwell

Registered Office
John Wiley & Sons, Ltd, The Atrium, Southern Gate, Chichester, West Sussex, PO19 8SQ, UK

Editorial Offices
350 Main Street, Malden, MA 02148-5020, USA
9600 Garsington Road, Oxford, OX4 2DQ, UK
The Atrium, Southern Gate, Chichester, West Sussex, PO19 8SQ, UK

For details of our global editorial offices, for customer services, and for information about how
to apply for permission to reuse the copyright material in this book please see our website at
www.wiley.com/wiley-blackwell.

Library of Congress Cataloging-in-Publication Data

Vasudevan, Alexander.
  Metropolitan preoccupations : the spatial politics of squatting in Berlin / Alexander Vasudevan.
    pages   cm. – (RGS-IBG book series)
  Includes bibliographical references and index.
  ISBN 978-1-118-75059-9 (cloth) – ISBN 978-1-118-75060-5 (pbk.)   1. Squatter settlements–
Germany–Berlin–History.   2. Squatters–Political activity–Germany–Berlin–History
3. Housing–Germany–Berlin–History.   4. Protest movements–Germany–Berlin–History–20th
century.   5. Human geography–Political aspects.   I. Title.
  HD7287.96.G32B483 2016
  307.3′36–dc23
                                    2015019948
A catalogue record for this book is available from the British Library.

Cover image: Squatted House with banner ("It is better that our young people occupy
            empty houses rather than foreign countries"). Richardplatz, Berlin-Neukölln,
            June 8, 1982. (Michael Kipp/Umbruch Bildarchiv)

Set in 10/12pt Plantin by SPi Global, Pondicherry, India

The information, practices and views in this book are those of the author(s) and do not necessarily
reflect the opinion of the Royal Geographical Society (with IBG).

Printed and bound in Malaysia by Vivar Printing Sdn Bhd

1   2015

For Megan

Contents

Series Editors' Preface

The RGS-IBG Book Series only publishes work of the highest international standing. Its emphasis is on distinctive new developments in human and physical geography, although it is also open to contributions from cognate disciplines whose interests overlap with those of geographers. The Series places strong emphasis on theoretically-informed and empirically-strong texts. Reflecting the vibrant and diverse theoretical and empirical agendas that characterize the contemporary discipline, contributions are expected to inform, challenge and stimulate the reader. Overall, the RGS-IBG Book Series seeks to promote scholarly publications that leave an intellectual mark and change the way readers think about particular issues, methods or theories.

For details on how to submit a proposal please visit:
www.rgsbookseries.com

David Featherstone
University of Glasgow, UK

Tim Allott
University of Manchester, UK

RGS-IBG Book Series Editors

List of Figures

Acknowledgements

The origins of this book can be traced to a bus tour of West Berlin that I was taken on by my mum in the summer of 1987 as a young child. While the tour took in all the expected sites, it also wound its way through the streets of Kreuzberg, a neighbourhood still reeling from the violent clashes between police and protesters on May Day and the subsequent police blockade which sealed off the district during the visit of American President Ronald Reagan. I was captivated, in particular, by the neighbourhood's recent history as an alternative enclave in which artists, dropouts, migrants, students, and workers all made a home. I realise now that this is a powerful imaginary with its own biases and blind spots. I was, nevertheless, struck by the graffiti and the colourful exterior of many apartment blocks in the neighbourhood. My mum explained to me that these were houses that had been occupied in the early 1980s by squatters who lived in them without paying any rent and that some had been legalised while, in the case of many other houses, the residents had been evicted.

I was immediately interested in who these 'squatters' were and why they chose to live illegally and yet, as it seemed to me at the time, in a radically different city. It would take me many years to finally formulate a project in which I was able to return to these questions. I began research on the history and geography of squatting in Berlin in 2006 and my numerous trips to Berlin were generously supported by funding at the University of Nottingham (New Researcher Fund and School of Geography Research funding) and a Small Research Grant from the British Academy (RA 1555). This is a project that has also benefited from various conference presentations and research seminars in the United Kingdom, the United States, Germany and Sweden. The debts to my hosts and the various conversations that emerged out of these presentations are too numerous to do justice here and I simply want to register how immensely formative they have been.

Neil Coe has been a patient and supportive editor and I am indebted to him for his guidance throughout the writing process, while Jacqueline Scott at

Wiley-Blackwell has been an immense source of assistance in bringing this book into its current form. This is, in turn, a book that would have not been possible without the tremendous generosity of the many archivists with whom I worked across Berlin and Germany. I would especially like to thank Ulrike Groß at the APO-Archiv at the Free University for her assistance and hospitality, Reinhart Schwartz at the Hamburger Institut für Sozialforschung, the entire team at the Kreuzberg Museum, Thomas at the Papiertiger Archiv and, finally, Hermann Bach at the Umbruch Archiv. These are collections that mean a great deal to me and their survival and use are vital for how we might still come to think about and inhabit cities differently. Over the past eight years, I have spent many months in Berlin and this is a project that also grew out of the various conversations, discussions, and interviews that I conducted with activists in the city. Massive thanks go out to Andy Wolff and Christine Ziegler at the Regenbogenfabrik for their help and support with my research. I was also honoured to be invited to speak at the 30th anniversary of the Regenbogenfabrik in 2011 and be part of such an important event. More generally, this book would not have possible without the actions undertaken by squatters themselves. I am always reminded of a graffitied quote by Nietzsche from *Thus Spoke Zarathustra* that figured on the wall of a squat in Freiburg: "One must still have chaos in oneself to give birth to a dancing star (*man muss noch Chaos in sich haben, um einen tanzenden Stern gebären zu können*)." I can't help but think that cities would become more just and liveable places if we were more willing to accept this sense of creative disorder (see Sennett, 1970).

I have also been very lucky to find myself within a wider academic community whose support and friendship has played a key role in this project. I would like, in particular, to thank my former supervisor Derek Gregory, whose continued support, advice and encouragement has been absolutely central to my development as a geographer. Heartfelt thanks also go out to Ben Anderson, Katherine Brickell, Gavin Brown, Stuart Elden, Matt Farish, Dave Featherstone, Melissa Fernandez, Matt Hannah, Jane Jacobs, Tariq Jazeel, Heike Jöns, Hayden Lorimer, Fraser MacDonald, Peter Merriman, Eric Olund, Chris Philo, David Pinder, Tom Slater, Quinn Slobodian, and Karen Till. I would also like to single out Colin McFarlane and Alex Jeffrey with whom I've collaborated over the past decade on a number of projects, who I am very lucky to count as both colleagues and friends, and whose work has played a key role in shaping this book. This is a project that has also taken me into a wider activist community whose work has been inspirational to me and who have challenged me in all sorts of ways. A big thank-you to Alex Andrews, Petra Davis, Gloria Dawson, Mara Ferreri, Dan Hancox, Huw Lemmey, Aaron Bastani, SQUASH and to all the students who occupied and protested in 2010 against the baleful marketisation of higher education. Those were actions that really mattered (*and still matter*). In Nottingham, this is a project that would not have been possible without my many terrific colleagues in the School of Geography. Thank you, in particular, to Stephen Daniels, Georgina Endfield, Isla Forsyth, Mike Heffernan, Lowri Jones, Dave Matless, Susanne Seymour,

Matthew Smallman-Raynor, Adam Swain, Lucy Veale, and especially Shaun French, Stephen Legg and Charles Watkins for encouraging me, inspiring me and challenging me. I have also learned a great deal and have enjoyed supervising Cordelia Freeman, Florence Todd-Fordham, Kevin Milburn, Hannah Neate, Jenny Rich and Charlie Veal and can only apologise for my endless reports about 'the book'. I would like to thank Elaine Watts for her assistance with the maps that appear in this book. Living in Nottingham has been so much easier knowing Alex Baker, David Bell, Emily Bennett, Cat Gold, Aaron Juneau and Marie Thompson.

Finally, this book would not have been possible without the support of family. To Christine and David McFarlane for their warmth and support and for always making me feel at home. To Douglas McFarlane and Daisy Mallabar for putting up with my endless chatter about activists and housing politics and for putting me up on my various travels through London. To my parents for their unwavering love and support and to my mum, in particular, for helping me with countless questions about translation. To Robert and Kläre Schölz for encouraging me to remember that the struggle for social justice is why I do what I do. And to Megan whose enthusiasm for this project has meant everything, and whose love and patience are a reminder of what really matters.

Chapters 4 and 6 draw in part on materials that first appeared in *Social and Cultural Geography* as "Dramaturgies of Dissent: The Spatial Politics of Squatting in Berlin, 1968 – " (Vasudevan, 2011a) and appear here with permission of Taylor & Francis. The book builds on and extends argument recently outlined in two articles in *Progress in Human Geography* ("The Autonomous City: Towards a Critical Geography of Occupation" and "The Makeshift City: Towards a Global Geography of Squatting").

Chapter One
Introduction: Making Radical Urban Politics

Considering there are houses standing empty,
While you leave us homeless on the street,
We've decided that we're going to move in now,
We're tired of having nowhere dry to sleep.

Considering you will then
Threaten us with cannons and with guns,
We've now decided to fear
A bad life more than death.

Bertolt Brecht (1967: 655)[1]

We don't need any landowners because the houses belong to us.

Ton Steine Scherben (1972)[2]

On the evening of 1 May 1970, a small theatre troupe began an impromptu performance in the middle of a shopping district in a newly-built satellite city on the northern outskirts of West Berlin. The troupe, Hoffmann's Comic Teater, was a radical theatre ensemble formed in 1969 by three brothers, Gert, Peter and Ralph Möbius, at the height of the countercultural 'revolution' in West Germany. Wearing colourful costumes and masks and accompanied by a live band, they soon developed a reputation for staging politically daring events that took place in the streets of West Berlin and in the city's many youth homes (Brown, 2013:

Metropolitan Preoccupations: The Spatial Politics of Squatting in Berlin, First Edition.
Alexander Vasudevan.
© 2015 John Wiley & Sons, Ltd. Published 2015 by John Wiley & Sons, Ltd.

172; see Sichtermann, Johler, Stahl, 2000; Seidel, 2006). The performances focused, in particular, on the everyday conflicts that shaped the lives of Berlin's working-class residents. Audience participation was actively encouraged by the troupe who developed an engaged agitprop style in which "the predominant cultural and political consciousness of the audience member" became the "starting point for the planning and realisation of the play" (quoted in Brown, 2013: 173). Scenes were improvised while spectators were invited onto the 'stage' to act out scenes from their own lives.

On 1 May 1970, the troupe travelled to the Märkisches Viertel, a large modernist housing estate in the district of Reinickendorf whose construction was part of West Berlin's First Urban Renewal Programme initiated by then Mayor Willi Brandt in 1963. The programme was responsible for the widespread demolition of inner-city tenements and the 'decanting' of their predominantly working-class occupants – approximately 140,000 Berliners – to new tower block estates on the fringes of the city (see Pugh, 2014; Urban, 2013). The performance by Hoffmann's Comic Teater focused, unsurprisingly, on the experience of the estate's residents and their anger at the lack of social infrastructure and the unwillingness of state-operated landowner and developer GESOBAU to provide "free spaces (*Freizeiträumen*)" for local youth.[3] It concluded with a scene that dramatised the recent closure of an after-school club (*Schülerladen*) after which the participants and spectators were encouraged to occupy a nearby building as a symbolic protest against GESOBAU. They were prevented from doing so, however, by the police who had been following the performance and had already secured the site. A group of over one hundred activists, performers and other local residents were nevertheless able to stage an occupation in an adjoining factory. As they began discussions over the formation of an autonomous self-organised youth centre, the factory hall was stormed by riot police and the occupiers, who included the journalist Ulrike Meinhof, brutally evicted. Three protesters were seriously injured and taken to hospital (see Figure 1.1).[4]

In the immediate aftermath of the eviction, a small group of local activists initiated a discussion about the future direction of political mobilising in the Märkisches Viertel. A strategy paper was produced and circulated by the group who criticised the new housing estate and its developers for their insufficient attention to the needs and desires of its tenants (Beck et al., 1975). One of the authors of that unpublished paper was Meinhof, who only two weeks later would take part in the breakout of Andreas Baader from the reading room of the Social Studies Institute of West Berlin's Free University (*Freie Universität*), an event which led to the formation of the Red Army Faction (*Rote Armee Faktion* or RAF) (Aust, 1985).[5] Hoffmann's Comic Teater continued to produce engaged performances in the wake of the occupation and also turned their attention to children's theatre (see Möbius, 1973). Members of the group were later involved in the formation of Ton Steine Scherben, one of the most important bands within the radical scene in West Berlin and whose history is largely inseparable from the

Figure 1.1   Arrest of the journalist Ulrike Meinhof at a protest occupation in the Märkisches Viertel in West Berlin, 1 May 1970 (Klaus Mehner, BerlinPressServices).

evolution of the anti-authoritarian Left in the city (Brown, 2009). While the factory occupation in the Märkisches Viertel was itself short-lived, it was nevertheless the first squatted space in a city where the radical politics of occupation would soon assume a new and enduring significance.

The story behind Berlin's first squat brings together a number of themes that are at the heart of this book: namely, the turn to squatting and occupation-based practices, more generally, as part of the repertoire of contentious performances adopted by activists, students, workers and other local residents across West Germany during the anti-authoritarian revolt of the 1960s and 1970s and in its wake; the relationship between the emergence of the New Left in West Germany and the transformation of Berlin into a veritable theatre of dissent, protest and resistance (see Davis, 2008); the recognition of uneven development and housing inequality as a source of political mobilisation and the concomitant privileging of concrete local struggles in Berlin for the composition of new spaces of action, self-determination and solidarity; and, finally, the widespread desire to reimagine and live the city differently and to reclaim a 'politics of habitation' and an alternative 'right to a city' shaped by new intersections and possibilities (Lefebvre, 2014, 1996; see also Simone, 2014; Vasudevan, 2011a; Vasudevan, 2014a).[6]

In the pages that follow, I develop a close reading of the history of squatting in Berlin. To do so, the book charts the everyday spatial practices and political imaginaries of squatters. It examines the assembling of alternative collective spaces in the city of Berlin and takes in developments in both former West and East Berlin. For squatters, the city of Berlin came to represent both a site of political protest and creative re-appropriation. The central aim of the study is to show how the history of squatting in Berlin formed part of a broader narrative of urban development, dispossession and resistance. It draws particular attention to the ways in which squatting and other occupation-based practices re-imagined the city as a space of refuge, gathering and subversion. This reflects the fulsome emergence of new social movements in the 1960s and 1970s in West Germany as well as the tentative development of an alternative public sphere in the final years of the German Democratic Republic (see Brown, 2013; Davis, 2008; Klimke, 2010; Moldt, 2005, 2008; Reichardt, 2014; Thomas, 2003). At the same time, it is a story that speaks to a renewed form of emancipatory urban politics and the possibility of forging new ways of thinking about and inhabiting the city that extend well beyond Berlin and, for that matter, Germany.

As the first book-length study of the cultural and political geographies of squatting in Berlin, this is a project that seeks to develop a rich historical account of the various struggles in the city over the *making of an alternative urban imagination* and the search for new radical solutions to a lack of housing and infrastructure. The book focuses, in particular, on what squatters actually *did*, the terms and tactics they deployed, the ideas and spaces they created. This is a history, in turn, that has had a significant impact on the transformation of Berlin's urban landscape and has shaped recent struggles over the city's identity. As I argue, squatters and the spaces they occupied were never incidental minor details in the formation and evolution of the New Left in West Germany in the 1960s and the various social movements which developed in the decades that followed. They played, if anything, a vital role in opening up new perspectives on the very form and substance that radical political action and solidarity could assume and are supported, in turn, by figures that point to an alternative milieu made up of thousands of activists and an even larger circle of sympathisers (Amantine, 2012; Azozomox, 2014a, n.d; see also Reichardt, 2014 for a wider perspective).

In Berlin, there have been at least 610 separate squats of a broadly political nature between 1970 and 2014 (see Figures 1.2 and 1.3). The majority of these actions took place in the city's old tenement blocks although they also encompassed a range of other sites from abandoned villas, factories and schools, to parks, vacant plots and even, in one case, a part of the 'death strip' that formed the border between West and East Berlin. As a form of illegal occupation, squatting typically fell under §123 of the German Criminal Code ("Trespassing") though many magistrates in Berlin as well as elsewhere in West Germany were reluctant to charge squatters as, in their eyes, a run-down apartment did not satisfy the legal test for an apartment or a "pacified estate (*befriedetes Besitztum*)" (Schön, 1982).[7] There were, in this context, two major waves of squatting in the city. The first wave between 1979 and 1984 involved 265 separate sites as activists

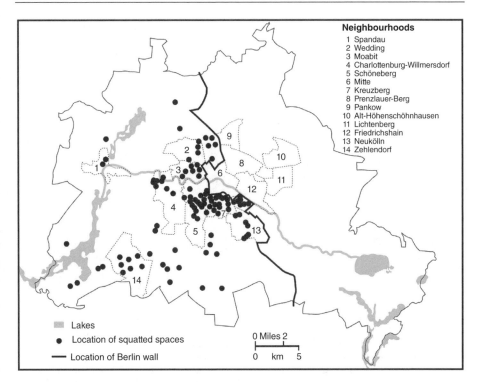

Figure 1.2 Map of squatted spaces in West Berlin up to the end of 1981. Map produced by Elaine Watts, University of Nottingham.

and other local residents responded to a deepening housing crisis by occupying apartments, the overwhelming majority of which were located in the districts of Kreuzberg and Schöneberg. At the high point of this wave in the spring of 1981, it is estimated that there were at least 2000 active squatters in West Berlin and tens of thousands of supporters (Reichardt, 2014: 519). The second wave between 1989 and 1990 shifted the gravity of the scene to the former East as hundreds of activists exploited the political power vacuum that accompanied the fall of the Berlin Wall, squatting 183 sites both in the former East as well as the West.[8] Since 1991, there have been only 100 occupations across Berlin as local authorities have vigorously proscribed and neutralised attempts to squat. Of these squats, 56 were evicted by the police within four days. Overall, 200 spaces have been legalised and, in 35 cases, the squatters have themselves acquired ownership (see Azozomox, n.d.).[9] While these figures point to the sheer scale and intensity of squatting in Berlin, they do not take into account other forms of deprivation-based squatting carried out by homeless people nor do they include the large number of East Berliners who, from the late 1960s to the end of GDR, illegally

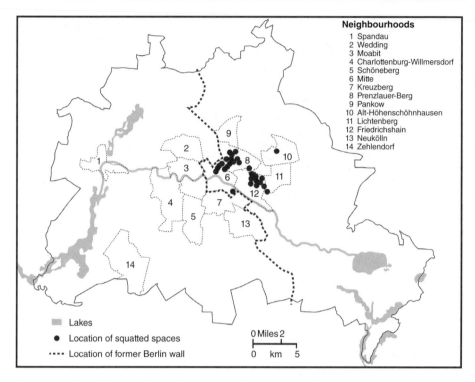

Figure 1.3   Map of the second wave of squatting in the former East of Berlin, 1989–1990. Map produced by Elaine Watts, University of Nottingham.

occupied empty flats in response to basic housing needs, a process that was known as '*Schwarzwohnen*' (Grashoff, 2011a, 2011b; Vasudevan, 2013).[10]

As these figures suggest, the history of squatting in Berlin occupied a significant place within a complex landscape of protest in the city. At the same time, the squatter 'movement' that emerged in Berlin was also connected to similar scenes in other West German cities in the 1970s and 1980s – most notably Frankfurt, Freiburg and Hamburg – and to a number of cities in the former East in the early 1990s (Dresden, Halle, Leipzig and Potsdam) (see Amantine, 2012; Dellwo and Baer, 2012, 2013; Grashoff, 2011b). It is perhaps surprising, therefore, that there remains little empirical work on the role of squatting – and the built form and geography more generally – in the creation and circulation of new activist imaginations and the production of collective modes of living. Why, in other words, did thousands of activists and citizens choose to break the law and occupy empty flats and other buildings across Germany and Berlin, in particular? Were these actions dictated by pure necessity or did they represent a newfound desire to imagine other ways of living together? Who were these squatters? What were the central characteristics of urban squatting (goals, action repertoires, political influences)? And in what way did these practices promote an alternative vision of the city as a key site of "political action and revolt" (Harvey, 2012: 118–119)?

In order to answer these questions, the study develops a conceptually rigorous and empirically grounded approach to the emergence of squatting in Berlin. More specifically, it develops three interrelated perspectives on the everyday practices of squatters in the city and their relationship to recent debates about the 'right to the city' and the potential for composing other critical urbanisms (see Attoh, 2011; Harvey, 2008, 2012; Lefebvre, 1996; Mitchell, 2003; Nicholls, 2008; Purcell, 2003; Vasudevan, 2014a). Firstly, it signals a challenge to existing historical scholarship on the New Left in Germany by arguing that the time has come to *spatialise* the events, practices and participants that shaped the history of the anti-authoritarian revolt and to retrace the complex geographies of connection and solidarity that were at its heart. Secondly, it draws attention to squatted spaces as alternative sites of habitation, that speak to a radically different sense of 'cityness', i.e. a city's capacity to continuously reorganise and structure the ways in which people, places, materials and ideas come together (Simone, 2010, 2014). Thirdly, it places particular emphasis on the material processes – experimental, makeshift and precarious – through which squatters came together as a social movement, sometimes successfully, sometimes less so. At stake here is a critical understanding and detailed examination of the conceptual resources and empirical domains through which an alternative right to the city is articulated, lived and contested (McFarlane, 2011b). A large part of this effort is, in turn, predicated on identifying concrete ways to recognise and represent the various efforts of squatters whilst acknowledging their complexity, contradictions, successes, and failures (see Simone, 2014: xi). To do so, the book ultimately argues, is to also draw wider lessons for how we, as geographers and urbanists, come to understand the city as a site of political contestation.

Spatialising the Anti-Authoritarian Revolt

In recent years, the historical development of the New Left in West Germany has become a growing area of scholarly activity as a new wave of studies have challenged the ways in which the West German student movement and its various afterlives have been narrated. Traditionally, the era known as '68 has been framed as "the moment when West Germany began to earn its place among the Western democracies" (Slobodian, 2012: 5). According to this view, 1968 and the protests and struggles that emerged in its wake were widely seen as a key watershed event in the democratisation of West Germany. This is a story in which young West Germans rebelled against the "stifling atmosphere of cultural conformity" that shaped the immediate post-war period. In so doing, they challenged the hysteria of the Cold War whilst confronting their parents about the crimes of the Nazi past. For the historian Timothy Brown, "such demands [...] acquired a special potency in a West Germany poised precipitously on the front line of the Cold War and struggling with the legacy of a recent past marked by fascism, war and genocide" (2013: 4). The consensus view is that the actions of the '68ers' helped propel West Germany into an era of liberalisation which, in turn, provided the necessary conditions for a vibrant democratic society.

As I demonstrate in this book, this is an argument that works to polish up, obscure and eviscerate other political developments and radical trajectories within the New Left that exceeded simple categorisation and containment. The reduction of the West German '1968' to a single overarching narrative thus foreclosed any meaningful attempt to assess and interrogate its nature and legacy. It was, however, the very surplus of such an event, its ability to disrupt existing explanatory models, that ultimately led, as Kristin Ross (2002) has argued in a related context, to its de-historicisation and de-politicisation. Not only were the motivation and goals of the events' myriad actors (students, workers, apprentices, artists and many other citizens), erased but the complex multilayered causes and consequences of their actions conspicuously ignored. This tendency has, if anything, been reinforced by an "overrepresentation, among historians of the events, of veterans of the student movement, whose lack of critical distance from events readily results in a mixing up of historical events and personal biographies" (Brown, 2013: 2; see Aly, 2008; Enzensberger, 2004; Koenen, 2001; Kunzelmann, 1998; Langhans, 2008). This should not, however, be seen as a simple case of historiographic revisionism but rather an act of *confiscation* through which the very richness and complexity of a mass movement is reduced to the "individual itineraries of a few so-called leaders, spokesmen, or representatives". Collective revolt is thereby "defanged" and recast as the jurisdiction and judgement of a small group of select 'personalities' (Ross, 2002: 4).

The story described in the pages of this book is deliberately set against these partisan tendencies and builds on an emergent body of work that seeks to historicise the anti-authoritarian protests that took hold in West Germany in the late 1960s as political struggles against various forms of oppression. Unsurprisingly, the events of the West German '1968' have, in recent years, received extensive treatment within the German literature (Fahlenbrach, 2002; Gilcher-Holtey, 1998; Klimke and Scharloth, 2007, Kraushaar, 2000; März, 2012; Reichardt and Siegfried, 2010; Scharloth, 2010; Siegfried, 2008). While the anglophone literature remains relatively small, some historians have nevertheless argued that the faultlines of a new interpretation can already be detected, one centred on the transnational and global dimensions of the uprisings that took place in West Germany (see especially Brown, 2013). My own view is that a "future consensus interpretation", as suggested by one prominent historian, runs the risk of substituting one historical orthodoxy for another (Brown, 2013: 3). Recent perspectives suggest, in contrast, a number of interconnecting themes that point to the sheer scale and diversity of opposition that grew out of the student protests in 1968. There has, in this context, been an attempt to pluralise the actors that were involved in the anti-authoritarian revolt and to argue that the New Left depended on the negotiation of gendered, classed and racialised moments of encounter and was, in fact, a product of participants from widely different backgrounds, orientations and experiences (Featherstone, 2012: 6; see Davis et al., 2010; Slobodian, 2012). Others have placed particular emphasis on re-thinking the protest movements of the 1960s and 1970s as a 'global phenomenon' that was a consequence of diverse translocal trajectories and connections (Höhn, 2008, Klimke, 2010; Slobodian, 2013a, 2013b, Tompkins, n.d.; see also Slobodian, 2012). Taken together, these approaches have shown that the construction of new movements and solidarities in West Germany was both an intensely local affair and one shaped by networks and relations that operated at a number of scales and which, in many cases, actively reshaped the terrain of political action.

Transnational histories of West German activists in the 1960s and 1970s have tended, as Quinn Slobodian has argued, to gravitate westwards and highlight the role of the United States in the development of the New Left in West Germany by retracing the exchange of protest repertoires and the movement of individuals across the Atlantic (2012: 6; see Klimke, 2010; Juchler, 1996; Höhn, 2008). While this work has yielded important insights into the entanglements between German and American oppositional cultures, it has also tended to obscure other alternative alliances and connections and downplay the impact of foreign students in drawing their West German counterparts into wider anti-imperialist struggles and, in the eyes of some commentators, into increasingly militant actions. To be sure, the emergence of a New Left internationalism in West Germany was often driven by abstractions and projections that reinforced, even instrumentalised, a mode of engagement "based on a West German Self and a Third World Other" (Slobodian, 2012: 11). And yet, it also promoted new collaborations with Third World actors which restored their agency and place within a radical history that was resolutely translocal and, as such, marked by deeply uneven geographies.

Attempts to capture the 'globality of 1968' have also encouraged greater sensitivity to questions of *periodisation*. There has developed, on the one hand, a new tendency in the historiography to adopt an approach that identifies the students protests of '1968' as the culmination of the 'long sixties' and "the climax of various developments that had been set in motion due to the immense speed of the social and economic transformations after the Second World War" (Klimke, 2010: 2; see Marwick, 1998). Other scholars, on the other hand, have returned to earlier trajectories that linked the protests in West Germany in the late 1960s to the radicalisation of many students and other activists and the subsequent turn by a portion of the anti-authoritarian movement to revolutionary violence in the 1970s (Hanshew, 2012; Weinhauer, Requate and Haupt, 2006). If the events of the German Autumn in 1977 – the kidnapping and murder of the industrialist Hanns-Martin Schleyer, the unsuccessful hijacking of a Lufthansa jet and the mass suicide of Red Army Faction (RAF) inmates in the Stannheim prison – are often seen as marking the end of the New Left, a new body of work has also returned to the 1970s with a view to recovering other histories of activism, dissent and self-organisation that emerged in counterpoint to groups such as the RAF for whom violence was becoming the exclusive means of struggle (Arps, 2011; Baumann, Gehrig and Büchse 2011; März, 2012; Slobodian, 2013). This work has been characterised, in no small part, by a new commitment to showing how extra-parliamentary groups were able to forge oppositional geographies and alternative lifeworlds that eschewed the "leaden solidarity" that seemingly defined the ways in which such groups were compelled to either declare solidarity or distance themselves from the actions of their violent comrades (Negt, 1995: 289; Slobodian, 2013: 224).

New attempts have, therefore, been made to examine the protest landscape that emerged in the 1970s in the wake of the student movement and to document the underground histories that were responsible for the appearance of various Marxist-Leninist and/or Maoist cadre parties, the so-called *K-Gruppen*, as well as the emergence of 'rank and file groups' (*'Basisgruppen'*) that turned to local neighbourhoods and other institutions (school, factories, etc) as a source of new initiatives and solidarities (see Arps, 2011; Kuhn, 2005). A small group of studies have also begun to explore the emergence of migrant activism in West Germany in the 1970s and 1980s and the ways in which foreigners remained active participants in a range of social movements (Bojadžijev, 2008; Karakayali, 2000, 2009; Seibert, 2008; Slobodian, 2013). Recent books by Tim Brown (2013) and Sven Reichardt (2014) have, in contrast, adopted a broader plenary approach that sets out to map the vast growth of alternative practices, projects and infrastructure in the 1970s and 1980s. Reichardt's thousand page account, in particular, retraces the emergence of an alternative milieu in West Germany in all its forms (agricultural communes, alternative bookshops, pubs and other businesses, social centres, experimental schools, neighbourhood workshops, etc.) and is one of few works that draws attention to the multiple spaces that were brought into being by activists across the country. Indeed, both Brown and Reichardt are at pains to acknowledge the significant role that squatting and other occupation-based practices played in the history of the anti-authoritarian revolt, though their accounts ultimately rely on an understanding of geography that is largely descriptive (see also MacDougall, 2011a, 2011b).

It is against this backdrop that I argue that the recent historicisation of the New Left in West Germany would also benefit from a critical framework that examines its complex *spatialisation*. By placing the everyday practices of squatters at the heart of this book, I seek to develop a geographical reading of the West German New Left and the activities and solidarities which emerged in the decades that followed. As I have already suggested, the history of squatting remains, in many respects, a blind spot within the wider historiography. The small number of studies that have been published in German are largely the work of activist-historians and have tended to place particular emphasis on specific aspects of the squatting scene at the expense of detailed historical coverage or wider theoretical reflection (Dellwo and Baer, 2012; Kölling, 2008; Laurisch, 1981; for an exception see Amantine, 2011, 2012). If this work identifies the importance of squatting to the recent history of a number of cities in Germany (Berlin, Frankfurt, Freiburg and Hamburg), my own account is predicated on a genealogy that focuses on Berlin and the long history of squatting-based activism in the city. This is a choice guided by the city's status as a key site within a wider landscape of protest and dissent (Davis, 2008; Vasudevan, 2011a). This is, moreover, a choice that has prompted me to take a number of risks. Firstly, I have chosen to widen my sightlines beyond conventional periodisations and take in developments both before and after the fall of the Berlin Wall whilst locating the imaginaries of squatters within a much wider narrative of displacement and dispossession. My retelling both acknowledges the importance of the New Left to the repertoire of contention developed by squatters and the ways in which such configurations of dissensus and habitation were continuously made, unmade and remade. It is not, in other words, my intention to suggest that the practices mobilised by squatters in Berlin in the early 1970s were somehow homologous to the actions adopted by protesters in the 1980s and 1990s. Rather, I trace an expansive understanding of the anti-authoritarian revolt that focuses on what became known as the *Häuserkampf* ('the housing struggle') and the different ways in which a crisis of housing shaped by repeated cycles of creative destruction became a crisis of dwelling characterised by a desire to re-imagine the city as a space of autonomy and self-determination. Secondly, I have also chosen to take in developments in East Berlin and explore an alternative history of occupation that stretched from the late 1960s to the fall of the Wall and which has remained largely undocumented (for an exception see Grashoff, 2011a, 2011b; see also Vasudevan, 2013). Whilst the actions of 'squatters' in the East differed from those mobilised in the West, what was referred to as *Schwarzwohnen* nevertheless played an important role in the development of a dissident public sphere in the German Democratic Republic (GDR) in the 1970s and 1980s and the new wave of squatting that erupted in Berlin (and elsewhere) in the winter of 1989.

This is a book guided by a commitment to marking the relationship between a spatial history of the anti-authoritarian revolt in West Germany, the everyday geographies of squatting and the *making* of an alternative urbanism. More specifically, this is an account that treats political activity and the various actions of squatters as *spatially generative*. The conceptual tools deployed throughout the book have therefore emerged from a detailed engagement with current geographical research on the politics of cities and the nature and constitution of urban struggles (Blomley, 2010; Datta, 2012; Dikeç, 2007; Graham, 2010; Harvey, 2012; Iveson, 2007; McFarlane, 2011b; Miller and Nicholls, 2013; Nicholls, 2008). They also form part of a larger normative project on the enduring significance of the city as a site of radical social transformation. As a geographer, my aim is to contribute to a re-thinking of how an alternative urban politics is produced, lived and contested and a deeper theoretical and empirical understanding of the practices of squatters. In so doing, I hope to provide a series of orientations that help us to reclaim a radically different right to the city shaped by a constituent desire to assemble and invent other urban spaces. This is, in other words, both a book on the history of squatting in Berlin and a critical commentary on how we conceptualise the city geographically and politically.

The Squatted City

In the conclusion to his book on urban squatting, the investigative journalist Robert Neuwirth (2006) remarks on how "the world's squatters give some reality to Henri Lefebvre's loose concept of 'the right to the city'". "They are excluded so they take," he writes, "but they are not seizing an abstract right, they are taking an actual place: a place to lay their heads. This act – to challenge society's denial of place by taking one of your own – is an assertion of being in a world that routinely denies people the dignity and the validity inherent in a home" (2006: 311). For Neuwirth, the seizure of place by squatters is itself an exercise in place making: "squatters, by building their own homes, are creating their own world" (2006: 306). This process of "dwelling-through-construction", as Neuwirth shows, is a product of countless everyday acts of adjustment and assembly, negotiation and improvisation (McFarlane, 2011a: 656). The lived city of squatters is, after all, a city structured by the shifting inequities that have come to characterise contemporary urbanisation. More often than not, to squat is to give form to a basic need for housing and shelter.

While the majority of the world's squatters continue to live in the Global South, as Neuwirth and others have shown, the hidden history of squatting is a global history (see also M. Davis, 2006; Vasudevan, 2014b). This is a history of makeshift rural cottages, precarious and informal urban settlements, experimental housing initiatives and radical autonomous communities. It is a history shaped by a complex patchwork of customary beliefs and rights, the improvised use of materials and skills, and the development of emergent forms of dwelling, sociality and cooperation. For the anarchist and historian Colin Ward (2002), the place of the squatter in the history of housing is far more significant, therefore, than is usually realised, and it would be wrong to subsume or equate the act of squatting – be it in the Global North or South – with the term 'slum'. If the latter's pejorative connotations are well established, the former's connection to a complex range of practices merits further scrutiny (McFarlane, 2008; Pithouse, 2006; Roy, 2011). This is borne out by the rich and evocative nomenclature for squatted communities across the globe, from *favela* in Brazil to *barriadas* in Peru, from *kijiji* in Kenya to *jodpadpatti* in India (Ward, 2002; see also Neuwirth, 2006: 16). And this is to say nothing of the equally large vocabulary of occupation developed by housing activists across cities in Europe and North America as part of a wave of squatting that began in the late 1960s (Birke and Larsen, 2007; Owens, 2008; Péchu, 2010; SqEK, 2013, 2014; Van der Steen, Katzeff, and Van Hoogenhuijze, 2014; Vasudevan, 2011a; Waits and Wolmar, 1980).

Squatting can be defined, in these contexts, as "living in – or using otherwise – a dwelling without the consent of the owner. Squatters take buildings [or land] with the intention of relatively (>1 year) long-term use" (Pruijt, 2013: 19). Squatting, to be sure, represents only one example of the many different strategies of shelter adopted by the urban poor that include more formal options such as 'hand-me-down' housing, hostels and purpose-built tenements, as well as informal forms of settlement from 'pirated subdivisions' to irregular peri-urban townships and other zones of extreme biopolitical abandonment (see M. Davis, 2006; Biehl, 2005; Roy, 2011). Unsurprisingly, accurate statistics are difficult to come by as the number of urban squatters is often deliberately undercounted by officials. It is estimated that there are anywhere from 600 million to 1 billion people squatting globally, with the vast majority located in cities and towns in the Global South (M. Davis, 2006: 23; Neuwirth, 2006; Tannerfeldt and Ljung, 2006). Even the UN's own restrictive definition identifies at least 921 million slum dwellers in 2001, with the number rising to over a billion by 2005, a high percentage of whom are squatters (M. Davis, 2006: 23). Set against this backdrop, the squatting movements that emerged in cities in the Global North in the 1960s and 1970s were admittedly smaller in scale – numbering in the tens of thousands – although they still played a significant role in the development of new forms of grassroots urban politics.

In a recent set of papers, I identified a set of analytical frames that seek to imagine and inhabit the possibilities of conceiving, researching and writing a global geography of squatting (Vasudevan, 2014a, 2014b; see also McFarlane and Vasudevan, 2013). As I argued, an optic is now needed that seeks to work across the North-South divide whilst acknowledging the differing purchase that certain political-theoretical constructs can and should have in dealing with squatting in different places. It was not, in other words, my intention to develop a theory of occupation and resistance that is all-encompassing. Whether it is Berlin or Mumbai, London or Nairobi, for most squatters the struggle begins, as Pithouse (2006) has suggested, with *this* land, this eviction, this neighbourhood, this developer, this idea, these needs. What therefore matters are the *connectivities* across multiple sites and how we might link a practical concern for the everyday struggles of squatters with a set of theoretical propositions that seek to open up a problem space for rethinking what it means to "see like a city" (Amin, 2013). To do so demands, on the one hand, a greater commitment to thinking about different contingent histories of precarious city life and how they might be *shared* across the North/South divide as the basis for new research platforms. It also depends, on the other hand, on a critical perspective that zooms in on the spatial practices of squatters, the different resources and materials they mobilise, and the ideas, knowledges and spaces they produce.

My aim in this book is more focused. While it builds carefully on these earlier theoretical intercessions, its emphasis is on one city, Berlin, and on extensive fieldwork that I conducted there over an eight-year period into the history of squatting and other forms of housing activism. The main arc of the book's argument thus emerges from the field and its archival remainders and is rooted in the often precarious and uneven intersections between social life, material infrastructure and politics that have shaped the recent history of squatting in Berlin as well as elsewhere across Europe (Simone, 2014: 2). This is, moreover, a book squarely embedded within an historical geography of squatting in Europe that began in the late 1960s and has played an important role in the social and political life of cities including Berlin, Amsterdam, Copenhagen, Frankfurt, Hamburg, London, Milan and Rome. My aspiration here is to develop a form of *radical spatial history* in which the details of a particular city can generate significant concepts that lead to a better understanding of wider historical formations and which, in turn, serve as critical instruments for uncovering aspects of a city's own history that may otherwise be ignored or forgotten (Simone, 2014: 2, 21). This is an approach that follows the various geographies of action, connection and engagement that underpinned the historical development of squatting in Europe as an alternative urbanism. If it places particular emphasis on trajectories of activism and resistance whose origins are in the Global North, these are stories whose recovery and recounting also point to ways of thinking about urban life that have wider resonances for geographers as well as other scholars, citizens and activists living in an age of planetary urbanisation (Brenner and Schmid, 2011, 2014; Merrifield, 2013b, 2013c).

The book thus speaks to a small but growing body of scholarship on the veritable explosion of squatting in Europe in the 1970s and 1980s that grew first in countries like Denmark, the Netherlands, Germany, the UK, France, Switzerland and Italy and, in more recent decades, in places such as Spain, Greece and Poland (Bieri, 2012; Mikkelsen and Karpantschof, 2001; Owens, 2008; Pruijt, 2013; Van der Steen, Katzeff and Van Hoogenhuijze, 2014; Vasudevan, 2011a). For many, this wave of squatting represented a 'new urban movement' characterised by the development of practices around collective forms of self-determination, struggles against housing precarity and a broader commitment to alter-global concerns and extra-parliamentary modes of political engagement (López, 2013: 881). Research has focused, in particular, on the development of histories of occupation in specific cities and their relationship to wider logics of urban restructuring (Azozomox, 2014a; Birke and Larsen, 2007; Dellwo and Baer, 2013; Holm and Kuhn, 2010; Mudu, 2004; Owens, 2009; Suttner, 2011). At the same time, work has also clustered around a number of key themes from the various alternative identities and intimacies produced and performed in squatted spaces to the repertoire of contentious politics adopted and shared by squatters across a range of different sites (Azozomox, 2014a, Cook, 2013; Geronimo, 1992; Kadir, 2014; Katsiaficas, 2006). Others have zoomed in on the legal challenges posed by squatting, whilst the effects of an austerity urbanism have prompted some scholars to explore the re-emergence of squatting and other occupation-based practices as alternatives to the predations of contemporary capitalism (see the essays in SqEK, 2014; Vasudevan, 2011b, 2013, 2014a, 2014b).

Recent efforts by members of the Squatting Europe Kollective (SqEK) have revived, in this context, a vital form of scholar-activism that played an important role in documenting earlier histories of squatting across Europe (see Bailey, 1973; Laurisch, 1981; Waits and Wolmar, 1980; Ward, 2002). As a collective of activist-researchers closely linked to squatter scenes in a number of European cities, SqEK have developed a programme of research which seeks to "critically analyse the squatters' movement in its relevant contexts" and to connect squatters and activists to research practices that stress a "collaborative and dialogical approach to knowledge production". SqEK thus encourages theoretical and methodological approaches in which the "researcher is critically engaged in squatting" (SqEK, 2014: 19). This is, as they admit, a challenging and controversial issue that they believe will lead to greater reflexivity on the part of the researcher as well as an ethos of cooperation and horizontality that builds on experiences within squats. Whilst there is much to recommend in this view, there is also, it seems to me, a danger that it privileges a form of militant research in which immediate *proximity* becomes a unique marker and arbiter of understanding and commitment. It also runs the risk of promoting certain forms of scholarship as somehow *representative* of a wider movement even though the degree to which the practices of squatters across Europe were able to come together and cohere as a single recognisable urban movement remains open to debate. In this book, I develop a different approach to the history of squatting in Berlin that seeks to address the gap between 'official' genealogies of subversion and their actual forms of "elaboration, circulation, re-appropriation, resurgence". In order to do so, I have collected the various voices, ideas, practices and knowledges produced by squatters whilst retracing the "transversal paths of revolt" that they themselves often followed (Rancière et al. quoted in Ross, 2002: 128; see Rancière, 2012a). The book invites readers, in this way, to step in and think with and alongside squatters whose actions were documented in magazines, posters, films and other sources written and recorded in the white heat of the moment. It is not my intention, therefore, to produce a typology of squatting based on a differentiation of goals and motives as some scholars have attempted to do (see especially Pruijt, 2013). For many residents in Berlin, the very choice to illegally occupy a flat was predicated on a refusal to accept the categories and structures imposed on them. These were, in turn, abstractions that were often used by local authorities and the media as a means to divide squatters and foster tensions within a wider ecology of protest.[11] Much of my own effort has been taken up in developing a critical understanding of the everyday practices devised by squatters in Berlin that focuses on their emergent "world-making potentialities" (Muñoz, 2009: 56). As I show in the chapters that follow, the squat was a place of collective *world-making*; a place to imagine alternative worlds, to express anger and solidarity, to explore new identities and different intimacies, to experience and share new feelings, and to defy authority and live autonomously (Gould, 2009: 178; see Kanngieser, 2013). At stake here was the opportunity to build an alternative *habitus* where the very practice of 'occupation' became the basis for producing a different sense of shared city life.

The question that animates this book is ultimately this: in what way were the actions of squatters in Berlin constitutive of the city? How, in other words, were they able to build the necessary conditions – however fleeting – for the articulation of an emancipatory urban politics? Or to put it somewhat differently, to what extent were practices designed for the purposes of survival and the extension of often highly precarious forms of life able to offer a touchstone for other alternative imaginings of cityness (Pieterse, 2008: 14)? My own grappling with these concerns has prompted me to examine the different ways in which the squatting scene in Berlin was able to transform the urban landscape into a living archive of alternative knowledges, materials and resources. I am drawn in this respect to recent work on cities that explores the intricate intersections of people, practices, spaces and materials and how they serve as a basis for the making of common political forms (Silver, 2014; Simone, 2010, 2014). At the same time, I am equally indebted to approaches to the city that draw attention to the range of improvised tactics and coping strategies used by individuals and groups to "widen the possibilities of urban dwelling" (McFarlane, 2011b: 33; see McFarlane, 2011a; Pieterse, 2008; Simone, 2004). One major aspect of these various engagements is an understanding of city life that is radically open, uneven and shaped by momentary "gatherings of fragments, efforts and forces" (Simone, 2014: 4). Whilst this is both a productive and seductive view, especially for scholars working in Berlin, a city still routinely described (and sometimes condemned) as "always becoming and never being", it also unwittingly belies a presentism that tends to occlude the complex histories through which other livelihoods and platforms were established and took hold in the city.

Taking the lead of one of Berlin's former residents, the philosopher Walter Benjamin, I seek to develop a radical spatial history that insists on the need to accommodate the uncanny presence of the past in the present. For Benjamin, this demands a rigorous mode of historical writing in which the past is sharply counterposed with the present. As Benjamin tells us, "it is not that what is past casts its light on what is present, or what is present its light on what is past" (1999: 463, Konvulut N 3,1). Rather, it is a case, according to Benjamin, of recognising the way in which the two come together as 'constellations' that have the ability to interrupt a certain model of history, one bound inexorably to a narrative of progress and development and the lockstep march of its teleology (see Vasudevan, 2005). It is with Benjamin's own materialist methodology in mind that I offer a thick description of the spatial practices and material geographies mobilised by squatters in Berlin. My main impulses are, in other words, both historiographic and political, especially give recent planning attempts to normalise Berlin's built environment and provide the city with a single strong identity in which past moments of radical dissent and opposition are assigned a carefully choreographed role (Colomb, 2011; Till, 2005). If the actions of squatters, therefore, represent an important aspect of Berlin's recent past that merits further critical attention in its own right, they also point to the enduring potential for other ways of knowing and learning the city that may lead to "more socially just possibilities" (McFarlane,

2011b: 157).[12] I am not interested, therefore, in re-imaging the history of squatting in Berlin as a single unbroken narrative. Rather, I am motivated by a commitment to retracing a complex, fractured and uneven story of care and solidarity, dissent and disagreement and why it still matters for how geographers conceive of the city and its relationship to radical social movements and the everyday micropolitics of dwelling and resistance.

A radical historico-geographical approach to squatting thus highlights the relationship between squatting and broader ongoing struggles over the meaning of urban space. As the Marxist geographer David Harvey reminds us, cities have perhaps become the key site for a variety of spatial struggles which, for Harvey, speak to the "intimate connection between the development of capitalism and urbanization" (2008: 24; see Harvey, 2014). Squatting may plausibly be seen, in this respect, as the political *other* to 'creative destruction' such that we continue to find, in the practices of squatters and the spaces they produce, resources and tools for challenging and disrupting the disagreeable materialities of capitalist accumulation. It is argued that squatters in Berlin have always cultivated an ethos of self-determination and autonomy – a radical DIY empiricism – that focused on the rehabilitation of buildings and the active assembling of new forms of dwelling. In practical terms, this depended on a modest ontology of mending and repair as squatters often confronted abandoned spaces that required significant renovation (McFarlane and Vasudevan, 2013; see also Vasudevan, 2011a). Makeshift materials and do-it-yourself practices combined with the sharing of food and other resources to provide the material supports for collective self-management. Squatted spaces represented, in this way, a fragile combination of materials, ideas, knowledges and practices through which other identities and intimacies were performed and new commonalities and solidarities developed and shared. As sites of reappropriation and rehabilitation, squats offered a suitable arena for challenging the "capitalist production of urban space" whilst playing an important role in the constitution of wider infrastructures and networks that combined housing needs and desires with broader political actions and other closely-related campaigns practices (anti-fascist organising, migrants' and precarious workers' rights, urban gardening schemes, etc.) (López, 2013: 870, 875; see Chatterton and Hodkinson, 2006; Mudu, 2004). But, more than this, squats in Berlin were not only embedded within local oppositional geographies and knowledges, they were also increasingly dependent on a host of translocal connections that linked activists across northern and southern Europe and which played a crucial role in the circulation and assembling of an alternative urbanism (Vasudevan, 2014a; see also SqEK, 2013, 2014; Van der Steen, Katzeff, and Van Hoogenhuijze, 2014). In the end, a detailed examination of these sites and the relations between them presents an opportunity to invigorate analyses of radical politics in Germany with a new attention to geography and the built environment – which has been explicitly lacking in most studies of the New Left (for an exception, see Hannah, 2010) – and to highlight, in turn, the generative role of the city as a key locus of protest both within Germany and elsewhere.

Towards a Spatial Grammar of Squatting

In its detailed emphasis on the micropolitical tactics and inventive geographies produced by squatters, this is a book that points to the co-constitution of the urban and the political whilst drawing particular attention to the materials with which radical political spaces are assembled and shared and the encounters and practices through which they are stabilised. This is an approach that therefore departs from accounts of urban social movements that frame the city as simply a platform or arena in which injustices are contested and new alliances and solidarities produced. In many of these accounts, the urban only appears "as a means to an end rather than an end in its own right" (Miller and Nicholls, 2013: 453). More importantly, this is also an approach that detects, in the actions of squatters, a *generative urbanism* that is antagonistic and subversive, inventive and open-ended (see Merrifield, 2013b). To do so not only depends on a more provisional understanding of how urban social movements are constructed. It also accords them an active role in shaping the terms on which they are shaped. At stake here is a concern for engaging with urban geographies that are themselves productive (Featherstone, 2012: 39).

Perhaps more than anything else, this is ultimately a book about the city as a radical political project. Whilst this is a project that has prompted some commentators to jettison existing framings, most notably Henri Lefebvre's 'right to the city' as a political placeholder, my own view is that these characterisations tend to side-step Lefebvre's original intentions, intentions that were themselves a product of the uprisings of the late 1960s and early 1970s (Merrifield, 2013c; see Lefebvre, 1996 [1976]). At the heart of Lefebvre's project, after all, is an understanding of the city as a work – an *oeuvre* [*ouvre*] – produced by the daily actions of those who live in the city. The right to the city, according to Lefebvre, is a right to inhabitation, appropriation and participation. It is both the right to inhabit and *be in the city* and the right to redefine and produce the city in terms that challenge the routinising demands of capitalist accumulation. Lefebvre's rights are, in this way, "rights of use rather than rights of exchange" (Purcell, 2003: 578). The right to re-appropriation thus implies the right to reclaim and reconfigure urban space as an oeuvre and "to maximise use value for residents rather than to maximise exchange value for capital" (Purcell, 2003: 578).

Lefebvre's positive re-affirmation of a right to habitation engages the problem of necessity and precarity head-on. It also, in my view, allows us to retain a right to the city that is *open-ended* and responsive to a politics that is both prefigurative and nonrepresentational. For Lefebvre, such an articulation of a radical urban politics can also be extended to the concept of "*autogestion*" which he uses to describe a process of worker autonomy and self-management and which should, in his view, be extended beyond the factory and into all spheres of everyday life (the state, the family, education, etc.). "Each time," he writes, "a social group...

refuses to accept passively its conditions of existence, of life, of survival, each time such a group forces itself not only to understand but to master its own conditions of existence, *autogestion* is occurring" (Lefebvre, 2009: 135). The political project of *autogestion* is, in other words, a constitutively geographical project to "transform the way we produce and use space" (Purcell, 2013: 41). At stake here, Lefebvre argues, is the "production of a space that is other" (1991: 391). Lefebvre describes this space as a "differential space" whereby

> Living labour can produce something that is no longer a thing…needs and desires can reappear as such, informing both the act of producing and its products. There still exist – and there may exist in the future – spaces for play, spaces for enjoyment, architectures of wisdom or pleasure. In and by means of [differential] space, the work may shine through the product, use value may gain the upper hand over exchange value: appropriation…may (virtually) achieve domination over domination, as the imaginary and the utopian incorporate (or are incorporated) into the real… (1991: 348).

If Lefebvre's understanding of autogestion and differential space points to a different kind of politics – autonomous, common and prefigurative – (Purcell, 2013), it also foregrounds the importance of re-appropriating space for the production of a "transformed and renewed right to urban life" (Lefebvre, 1996: 158).

There is, of course, no doubt that a workable notion of the right to the city must still confront the contradictions, divisions and exclusions implicit in rights claims. At the same time, it is equally important to recognise the *constituent* dimension of Lefebvre's original claims. Whilst some squatters in Berlin were certainly familiar with the terms of Lefebvre's argument, many were not. This did not stop them, however, from articulating a right to the city that was also a right to housing and infrastructure, a right to free space and self-determination, a right to explore other identities and a basic right to be in the city. What ultimately mattered, more than anything else, was a right to participate in the production of urban space and a desire to generate new counter-geographies of adaptation and experimentation, protest and dissent (Lefebvre, 1991; see Purcell, 2003). It is in this context that I therefore develop a *critical geography of occupation* as a political process that materialises the social order which it seeks to enact. Occupation, according to this view, involves different ways of extending bodies, objects and practices into space in order to create new alternative lifeworlds. The relationship between occupation-based practices of squatters in Berlin and the production of a renewed right to the city depended, in other words, on the mobilisation of an "experimental politics" (Lazzarato, 2009) and an "imaginary pragmatics" (Merrifield, 2013b) that actively prefigured a radically different sense of what it means to think about and inhabit the city.

If this book advances a geographical framework for examining the forms of political agency mobilised by squatters, it does so through a conceptual armature that seeks to capture the very provisionality of squatting. This is a book that combines a commitment to radical historical research with modest theoretical concerns. It connects traditional Marxist urban geography (Harvey, 2012, 2014; Merrifield, 2014; Smith, 1996) with key work on urban social movements (Castells, 1983; Mayer, 2009, 2013b; Melucci, 1980; Tilly, 2008; Touraine, 1981). It also seeks to bridge wider theoretical engagements on the city (McFarlane, 2011b; Merrifield, 2013b, 2013c; Simone, 2010, 2014) with a range of cultural theory and radical political philosophy (Foucault, 2011; Guattari, 2009; Krahl, 1971; Marcuse 1964; Negri, 2005). The book advances a grounded theoretical imaginary that attempts to explain and critically interrogate how emancipatory urban politics are made and shared. It develops, as I have already argued, a close reading of the makeshift practices and experimental performances mobilised by squatters in direct opposition to inequality and oppression. It also highlights the relationship between the active assembling of urban 'infrastructures' and the production of situated connections and solidarities that linked squatters in Berlin to a range of activists, movements and practices. This was (and remains) a deeply fraught process and the book also confronts "some of the 'dark sides' of solidarity" and the forms of emotional labour that were both central in holding the scene together and in accentuating its divisions (Featherstone, 2012: 12; see Gould, 2009).

The book consists of five substantive chapters followed by a conclusion. It is broadly chronological in format and locates the genesis of the squatting movement in Berlin within a broader history of capitalist accumulation, creative destruction and uneven urban development. In Chapter 2, I attempt to position the relationship between squatting and a politics of housing in Berlin within a wider context of dispossession and resistance, and to provide a supporting framework for understanding the complex and uneven set of conditions that contributed to the emergence of Berlin's squatting scene. The chapter retraces the relationship between the urbanisation of capital and the long history of housing *precarity* in Berlin from the middle of the 19th century to the 1960s. It does so by showing how housing inequality in Berlin has depended on recurring cycles of creative destruction that repeatedly condemned significant numbers of people to misery and prompted many to seek informal forms of housing and shelter. As crises of capital became crises of *dwelling*, they also catalysed new forms of contentious politics. The chapter thus sets out to describe a complex history of adaptation and improvisation through which an alternative spatial politics was developed in Berlin. This is done in three ways. The chapter first considers the relationship between the extension plan for the city of Berlin drawn up by James Hobrecht in 1862 and the building of squatter settlements on the outskirts of the city in the 1860s and 1870s. It then shifts attention to widespread strikes over rising rents in the 1920s and 1930s. The final section of the chapter examines the evolution of new forms of protest in Berlin that emerged in response to the redevelopment of the city. In this way, the chapter retraces a series of spatial practices that shaped an alternative history of housing in Berlin. This is a history, it is argued, of protest, resistance and occupation that not only reclaimed space but transformed the built form into an instrument of resistance and creative re-appropriation.

Chapter 3 documents the emergence of the early squatting movement in West Berlin in the late 1960s and early 1970s. It traces the development of the squatter scene (*Besetzerszene*) through the changing set of tactics and practices adopted by the student movement and the extra-parliamentary opposition (*Außerparlamentarische Opposition* or APO) in the late 1960s. The chapter shows how a set of performances mobilised by various elements of the student movement and the APO in the 1960s – from happenings and teach-ins to new forms of theatre and agitprop – provided an important action repertoire that would later come to influence the spatial practices of squatters in both West and former East Berlin. Whilst these 'direct action' tactics blurred the traditional boundaries between theatrical performance and public space, they also assumed new forms in the wake of emergency laws that banned public political demonstrations in West Germany in the late 1960s. The chapter thus examines a range of oppositional geographies in West Berlin that were produced during this period from early experiments in alternative forms of communal living in West Berlin (Kommune I, Die Bülow- Kommune, Die Anarsch-Kommune) to the first squatted spaces (Georg von Rauch-Haus, Tommy Weisbecker-Haus). At the same time, the chapter situates the emergence of alternative forms of collective living and other self-organised projects as part of a broader turn to the emotional and material geographies of everyday life. Intimate settings – cafés, pubs, alternative presses, bookstores, youth centres, and squatted spaces – offered, it argues, a radical infrastructure through which alternative support networks were created, friendships made and solidarities secured. At stake here, as the chapter concludes, was the cultivation of political spaces and collective practices that promoted an alternative vision of the city.

Chapter 4 focuses on the period between 1979 and 1984 which represented the high point for the squatting movement in West Berlin. The chapter examines the practice of squatting and 'occupation' as an act of collective *world-making* through which an alternative understanding of shared city life was (quite literally) constructed. It concentrates, in particular, on how squatted spaces were assembled and sustained on an everyday basis. The chapter shows how this depended in no small part on a politics of *adaptation, mending and repair* that served as a direct response to an endemic housing crisis characterised by top-down planning initiatives, rampant property speculation and local corruption. Squatters in West Berlin often confronted abandoned spaces that required significant renovation and Chapter 4 offers a thick description of the wide range of practices and tactics deployed by squatters as they challenged and were later compelled, in many cases, to accommodate existing property regimes. Whilst the main aim of the chapter is to examine the everyday geographies of squatters in Berlin as a radical *makeshift urbanism*, it also draws attention to the complex constellation of affects, emotions and feelings and the decisive role that they came to play in the social life of a squatted house. Here, it is argued that the activities of squatters was dependent on a form of *emotional labour* through which the boundaries of 'activism' and 'the political' were constantly made, unmade and remade. In practical terms, the chapter begins with a discussion of the TUNIX ('Do Nothing') conference that took place in West Berlin at the end of January 1978. The conference brought together activists from across Europe to explore new forms of organisation and resistance in the wake of statist repression and growing leftist violence. The TUNIX conference provided an important point of departure for the development of new activist geographies and the remainder of the chapter examines the consequential emergence of the squatting movement in West Berlin in three stages. The first zooms in on the key period between 1979 and the 'hot summer' of 1981 at which point over 165 houses were occupied across West Berlin. The second reconstructs the range of material and emotional geographies produced by the squatters, while the third focuses on the period after 1981 and the dissolution of the squatting scene through protracted negotiations with West Berlin authorities, the legalisation of some occupied houses and the 'pacification' of the more 'militant' elements of the movement through criminalisation and eviction. Whilst the chapter demonstrates how the everyday spatial practices of squatters in West Berlin represented, for some, an act of militant *antagonism* and *insurgency* it also shows that, for others, it constituted a delicate balancing act between existing political institutions and forms of radical citizenship. In this way, the chapter offers an opportunity to closely examine how activists responded to decline and failure, dissent and violence. Squatted spaces, it concludes, were both sites of liberation and possibility and sources of intense conflict and struggle.

Chapter 5 traces the emergence of a second major wave of squatting in Berlin in the former East of the city as activists took advantage of the political power vacuum that accompanied the fall of the Berlin Wall in 1989. The chapter builds on the arguments set out in Chapters 3 and 4 in order to explore the complex combination of formal and informal practices – from planning, policy and law to everyday practices of dwelling and infrastructure – that shaped the development of squatting and other occupation-based practices in East Berlin before the fall of the Berlin Wall and in its immediate aftermath. The chapter is divided into two main parts. The first part seeks to reconstruct the relatively unknown history of illegal occupation in East Berlin (*Schwarzwohnen*), its relationship to both the development of a *critical public sphere* in the German Democratic Republic (GDR) in the 1970s and 1980s and the new wave of squatting that erupted in the winter of 1989 and which was set against a rapidly-changing landscape of property. The second part of the chapter tracks the further intensification of the squatting scene and the extension of a repertoire of contentious politics that played an important role in the history of the squatting scene in West Berlin. It also follows the growing conflict between West and East German activists in the months leading up to official German reunification. The chapter thus places particular emphasis on the fall of the Berlin Wall which marked the beginning of a rapid process of 'spatial redefinition' for the former GDR. This offered, it argues, a rare opportunity for housing activists to create and experiment with radically new and autonomous spaces as much of East Berlin's 19th-century housing stock was never properly maintained and had, by the late 1980s, slipped into serious disrepair. The chapter concludes, in this context, by revisiting the series of events that led to the violent clearing of squatters on Mainzer Straße in November 1990 and the wider implications that the evictions had on housing-based activism in Berlin.

Chapter 6 focuses on the transformation of Berlin's squatting scene in the wake of the Mainzer Straße evictions in November 1990. The police crackdown marked the beginning of the end of the squatter movement in Berlin and the eventual re-orientation of the scene around a new set of experimental practices. If urban squatting in Berlin had its origins in an insurgent form of 'self-help', the chapter examines how squatting had become a major mechanism in the commodification of urban space as tactics of informal urban living were quickly transformed into new strategies for neo-liberal urban renewal especially in districts of former East Berlin. The chapter highlights the 'capture' and instrumentalisation of occupation-based practices by the state and the relationship of squatting to gentrification and other forms of urban restructuring. At the same time, the chapter also explores how the tactics and strategies of urban squatting were adapted and reworked by groups of activists and artists as well as a host of other organisations in Berlin. It shows how the history of squatting transformed the city of Berlin into a living archive of alternative knowledges, materials and resources. This radical 'archive' was, it is argued, central to more recent attempts at developing alternative modes of urban living. These have taken on a number of different forms and the main aim of Chapter 6 is to work through a variety of examples that illustrate the importance of reclaiming a renewed right to a different city. The chapter offers three interrelated perspectives. Firstly, it discusses the role of cultural experimentation and artistic practice in the development of new strategies for participatory architecture, community design and everyday dwelling. Secondly, it explores the extension of a politics of occupation and inhabitation that centred on the cultivation of other identities and intimacies. Thirdly, it considers the recent emergence of a new round of housing activism in Berlin that has come to challenge gentrification, dispossession and rising rents (new forms of citizen occupation, protest camps and eviction resistance networks). The chapter concludes by showing how the practice of urban squatting continues to offer conceptual tools and practical resources through which a more radical and socially just urbanism may be produced.

The Conclusion returns to the wider theoretical framework developed in the Introduction. It closes by identifying the implications of urban squatting for how we think, research and inhabit the city. If the book argues for the need to hold on to the *potential* of other alternative urbanisms, it does so by showing how the historical geography of squatting in Berlin has been a deeply contested project and one through which the very forms and imaginaries of a better city have been continuously made and remade (see McFarlane, 2011b). At stake here is a right to the city that is characterised by a *constituent* desire to participate in the production of alternative urban spaces. As the book ultimately argues, the *normative* demand for an 'alternative city' has become an increasingly pressing issue and the history of squatting in Berlin offers not only a spatial history of occupation and resistance but a set of tactics and strategies for how we might still come to know and live the city differently.

At the centre of the book, in the end, is a commitment to recording the various actions and words of Berlin's squatter community. This is a community that has devoted significant energy to *archiving* their own practices and representations and to documenting the development of oppositional political cultures in the city. If this points to a self-consciousness and reflexivity on the part of squatters, it has also prompted me to develop a form of historical geography that is attentive to the words of its subjects and which seeks to let the facts of their own thoughts speak for themselves. It is perhaps not surprising that, as a genealogy of radical urban politics, one of the book's most important points of reference remains Jacques Rancière's *Proletarian Nights* (2012a). The dreams and visions of Rancière's 19th-century workers in Paris were, of course, different to those of squatters in 20th-century Berlin. What was important for me, in this context, was not to impose a new theory of spatial politics *onto* the actions of squatters, but instead to try and change "the very look of the *material of theory* itself" (Rancière, 2012b; emphasis added). The modest theoretical focus of this book thus emerges from concepts and debates that were immanent to the practices of squatters, the materials they mobilised and the spaces they created. These are spatial practices that, in my view, enlarge the ground of the political and how we, as geographers, document and attend to the city as a site of contestation and resistance, solidarity and experimentation.

As a form of radical spatial history, this is a book that is rooted in the vast repertoire of archives that remain an important testament to an alternative Berlin, a version of the city that is increasingly under threat by recent redevelopment and regeneration. Whilst my account is predicated on a close and detailed reading of a vast range of archival materials (eye witness accounts, letters, pamphlets, magazines, flyers, maps, photographs, documentary footage), it also acknowledges that the very existence of these makeshift archives is shaped by a conviction that effective forms of activism and protest "must necessarily emerge from an historically grounded understanding of the activist past" (Ford and Smith, 2014). I have made the decision, therefore, to focus my attention on these historical remainders and to push the numerous interviews that I conducted with former squatters and other community activists further into the background.[13] This was a difficult decision to make but one based on a realisation that the often heated and partisan nature of those conversations risked opening old wounds and divisions whilst fostering renewed feelings of anger, betrayal and resentment. These are feelings that have already taken an enormous toll on individuals and the wider 'scene' and it seemed to me that other solidarities and connections were now urgently needed. If I have therefore written an angry book of sorts, it is based on a growing sense of indignation at the baleful neo-liberalisation of Berlin and a realisation that there is now little room for the makeshift spaces produced by squatters and the alternative urbanisms they presumed. Attempts to squat in the city are routinely suppressed by the police, whilst the forced displacement of the city's most vulnerable tenants has become a commonplace. And yet, the recent resurgence of housing-based activism in Berlin also points to an appetite for building *common political spaces* of care, cohabitation and solidarity that seek to reclaim a right to the city and offer an alternative to an intensifying revanchism (see essays in Holm, 2014b). The book should ultimately be read, therefore, as an archaeology of our present and as a powerful critique of the neo-liberal city. Whilst this a project informed by both a scholarly and political commitment, the book does not seek to provide a romantic gloss on urban squatting. If anything, it confronts and examines the fraught and uneven nature of squatting in Berlin and seeks to open up a critical space for exploring alternative political imaginaries and their conditions of possibility. This is, in other words, a book intimately connected to the history of one city, Berlin, but equally committed to the fostering of alternative research configurations that extend to other cities and the wider logics of displacement and dispossession that they, in many respects, share. The political challenge, it argues, is to counter the accumulation of capital with new ways of dwelling differently that "are produced and held in common" (McFarlane, 2011b: 157; see Hardt and Negri, 2009). To do so, as the history of squatting documented in these pages shows, is to reveal the conditions – the counterarchive of practices, sentiments, tactics and stories – that point to an alternative urbanism. And it is these living geographies that ultimately hold the promise for the development of a different, better city.

Notes

1   All translations in the text are the author's unless otherwise stated. An attempt has been made to translate texts as faithfully to the spirit of the original as possible.

2   Song from Ton Steine Scherben, "Die letzte Schlacht gewinnen wir" (LP, 1972).

3   Hamburger Institüt für Sozialforschung (hereafter HIS), *Rote Presse Korrespondenz*, "Polizeiterror in Märkisches Viertel", 2, 64 (1970): 11.

4   HIS, "Polizeiterror in Märkisches Viertel", 11.

5   The Märkisches Viertel remained an important point of reference for Meinhof and appeared in many of the communiqués published by the Red Army Faction, including the group's first statement in the radical journal *Agit 883* on 5 June 1970 (see Colvin, 2009).

6   Throughout the book, I have used the term 'the anti-authoritarian revolt' which I have borrowed from Lönnendonker et al. (2002) and Timothy Brown (2013) in order to encompass the wide range of practices adopted by the New Left in West Germany in

the 1960s and 1970s. Whilst I have also adopted the widely-used term 'extra-parliamentary opposition' (*Außerparlamentarische Opposition* or APO), I have specifically done so when discussing the (self-described) alliance of the West German student movement with other social movements in the late 1960s.

7　See *Der Spiegel*, 19.4.1982. As jurists and legal scholars have shown, challenges to §123 of the German Criminal Code ("Trespass" or *Hausfriedensbruch*) were contingent on an argument that showed how abandoned homes and properties no longer satisfied the legal conditions necessary as a dwelling or "pacified estate" (*befriedeten Besitztums*). The origins of the legal terms of reference date back to Prussian times and the late 18th century in the first instance (see Rampf, 2009).

8　Figures for the number of squatters involved in the first major wave of squatting in West Berlin in the late 1970s and early 1980s fluctuate anywhere from 1000 to 5000 (see Amantine, 2012: 18; Rosenbladt, 1981: 40). The political vacuum that accompanied the fall of the Berlin Wall in 1989 has made it difficult to produce similar data for the second wave between 1989 and 1990.

9　I've constructed a database of Berlin's squatted spaces using flyers, police press releases and other archival material from the Papiertiger Archiv in Berlin-Kreuzberg. I've cross-checked my data with the recently published website on Berlin's squatter movement (www.berlin-besetzt.de).

10　While it is difficult to reconstruct the full history of *Schwarzwohnen* in East Berlin, recent work by Udo Grashoff (2011b) suggests that there may have been thousands of illegal occcupiers living in flats across the city. I explore the topic in greater detail in Chapter 5.

11　I am drawing here, in particular, on the work of the French philosopher Jacques Rancière whose close reading (2012a) of the often forgotten writings of 19th-century workers in France represented a challenge to the identities that were usually conferred on them. These writings formed the basis of Rancière's book, *Proletarian Dreams*, but also informed his earlier work with the journal, *Révoltes Logiques* (see Ross, 2002).

12　Whilst I am sympathetic here to recent work that mobilises 'assemblage theory' as a way of attending to the realisation of alternative potentialities, I do worry that it presumes an understanding of history as a standing reserve for future mobilisations and alliances. There is danger that the sheer complexity of the past is elided in the rush to focus on the ways in which certain practices and knowledges are actualised (see McFarlane, 2011b).

13　Interviews were conducted between February 2008 and August 2013. For the sake of consistency, I have altered the names of all interviewees and used initials to ensure anonymity.

Chapter Two
Crisis and Critique

With housing you can just as easily kill someone as with an axe!
Heinrich Zille (quoted in Kowalczuk, 1992: 234)

Farewell and out we went,
we're moving without paying the rent!
Graffiti on Berlin door, October 1872 (quoted in Nitsche, 1981: 62)

In 1872, an article by the well-known author and feuilletonist Max Ring entitled "A Visit to Barackia" ("*Ein Besuch in Barackia*") was published in the popular German newspaper, *Die Gartenlaube*. Ring was a regular contributor to *Die Gartenlaube*, producing a series of vivid sketches on Berlin city life from the late 1850s to the mid 1870s. In "Ein Besuch", Ring recalls a conversation and city stroll with a friend (imaginary or otherwise) in what is now the district of Kreuzberg. The walk took them along the Tempelhofer Ufer, past the Hallesisches Tor to Kottbusser Damm, a journey through the new roads and *Gründerzeit* apartments that marked, so it seemed, Berlin's growing prosperity.[1] As they were walking, Ring's companion asked him whether "he had ever visited the Berlin Free Republic of Barackia?" "What republic? Is this one of your terrible jokes?" replied Ring. "Not at all," his companion insisted, "It is, in fact, a new free state in the proper sense of the word – both daring and significant – a state in the open air, in the open field, with the freest views and the most liberal institutions, free

Metropolitan Preoccupations: The Spatial Politics of Squatting in Berlin, First Edition.
Alexander Vasudevan.
© 2015 John Wiley & Sons, Ltd. Published 2015 by John Wiley & Sons, Ltd.

from all police repression, free from executors and tyrannical landlords, a place without rents and taxes, without stinking gutters and disreputable cesspools, free of all the expense and agony of the world city." "You're not talking about those unfortunate people," Ring objected, "who are unable to find any shelter because of the prevailing housing crisis?" "Unfortunate," his companion shot back, "these people are hardly unhappy [...] These poor people that live in their new 'barracks' are now [...] better and happier than they were ever before in their wretched attics and damp cellars, where they still feared an increase in rent or the prospect of a sudden eviction." As Ring's companion continued,

> I am pleased that on this occasion the world and imperial city shows itself in all its glory [...] There is money everywhere and the stock market no longer knows what to do with it. Every day, new banking associations are set up. Villas, houses, palaces, whole streets are growing like mushrooms out of the ground [...] Meanwhile there are thousands in Berlin who are now homeless and without shelter because they can no longer afford the continually increasing rent as a result of these abnormal conditions.

"It is the duty of the state, the city, society as a whole," he concluded, "to [build] cheap workers' dwellings [...] Since this has not been done, they have instinctively applied the principle of self-help (*das Princip der Selbsthülfe*) and thus embarked on their own" (Ring, 1872: 458).

As they were chatting, Ring and his companion turned southwards on Kottbusser Damm; they soon reached the Hasenheide Park and the open fields that stretched south of the city's official boundaries. At this point, they arrived at an encampment of "rough wooden huts", many of which were covered with cardboard sheets and woollen blankets. "We are now in the Republic of Barackia," declared Ring's friend. They were soon met by one of the encampment's inhabitants and were given a tour of the site. "Some of [the] cabins still betray a most primitive state," Ring noted, "however, the majority are not only comfortable, but are furnished and surrounded by petite plantings, vegetable patches and flower gardens, a friendly image of an improvised summer home in a wild and rural location." Almost all of the inhabitants, he added, "spoke with satisfaction about their present situation, and we heard many of them say that they have never had it better and that they wished to live here forever" (Ring, 1872: 460).[2] The remainder of the article is devoted to a further description of the encampment, one of a number of squatter settlements that sprang up in the early 1870s on the open fields outside the city gates that marked the official boundaries of Berlin (see Figure 2.1).[3] Whilst Ring's companion was at pains to identify the "abnormal conditions" that were, he believed, responsible for the establishment of these settlements, housing scarcity in Berlin was and remains an important aspect of the city's social and political geography. At stake here is a much deeper history of insecurity which has been an enduring reality for city residents and it is a detailed reconstruction of this history that is the main focus of this chapter. To do so, the

Figure 2.1 Berlin's 'shanty town'. Image in *Die Gartenlaube* by Ludwig Löffler (1819–1876), p. 459.

chapter examines how the recent history of housing inequality in Berlin is closely connected to the changing logics of capitalist urbanisation and a product of recurring cycles of creative destruction and accumulation by dispossession which have repeatedly condemned significant numbers of people to misery and prompted many to seek informal forms of housing and shelter (Harvey, 2008; 2012; see Hegemann, 1979 [1930]; Geist and Kürvers, 1980–89). The chapter thus focuses on how crises of capital became crises of *dwelling*. But, more than this, it sets out to describe a complex history of adaptation and improvisation through which many ordinary Berliners responded to what Friedrich Engels (1995 [1872]) once described as the "housing question". For Engels, this was a question that was never about housing *sensu stricto* but about the broader injustices produced by an overarching structure of socio-political interests that shaped struggles over property rights, accumulation strategies, and speculative land interests. Whilst Engels was writing with the shock cities of early industrialisation specifically in mind, the injustices of housing have only intensified over the past century or so, taking in cities in both the global North and South (see Harvey, 2008). It is in this context that this chapter zooms in on the recent history of Berlin with a view to forging an empirically grounded approach to the geographical study of housing and precarity. The chapter does so in three ways. It first considers the early history of housing scarcity in Berlin, focusing in particular on the relationship between the extension plan for the city of Berlin drawn up by James Hobrecht in 1862, riots over forced evictions in the 1860s and 1870s and the building of squatter settlements on the outskirts of the city during the same period (Bernet, 2004; Geist and Kürvers, 1980–89; Ladd, 1990). It then shifts attention to tenant organising and widespread strikes over rising rents in the 1920s and 1930s. The final section of the chapter examines the evolution of protests over urban redevelopment in the post-war era. Taken together, the chapter retraces a series of spatial practices that have come to shape an alternative history of housing in Berlin. This is a history, it is argued, of survival and necessity but also one of protest, resistance and occupation that transformed the built form into an instrument of self-help and re-appropriation.

It should come as no surprise, perhaps, that it was in fact some of Berlin's more recent housing activists – most notably squatters in the 1970s and 1980s – who first drew attention to the complex contours of this history and *their relationship to it* (see Kowalczuk, 1992; Nitsche, 1981). They undoubtedly did so in order to better understand the historically sedimented nature of their own struggles and the repertoire of tactics they deployed. But they also did so with a view to exploring the socio-economic dynamics which connected urban development in Berlin to *repeated crises of housing in the city*. While some activists and squatters took this opportunity to reduce an earlier history of housing-based activism in Berlin to a crude rehearsal of their own actions, at the heart of this chapter is a critical interrogation of the historical geography of capitalist urbanisation in Berlin. As I argue in what follows, the relationship between squatting and a wider politics of housing must be placed *within* a long history of dispossession and resistance rather than as the culmination of it. The chapter thus offers a thick description of the logics of creative destruction as a process that disrupts the "sedimentations and stabilities that inhere within the meanings, routines and expectations that usually attach to 'place'" (Gregory, 2006: 16). If this is a process that has prompted some geographers such as David Harvey to develop an impressive plenary account of the tense and turbulent landscape of capital, my own sightlines are more modest and closely allied to the particularities of dislocation and dispossession. At the heart of this chapter, in other words, is a commitment to registering an alternative history and geography of Berlin that is both a critical account of uneven geographical development and a story of protest and resistance, adaptation and invention.

The Making of the Tenement City

It was rapid urbanisation in the years after the founding of the German Reich in 1871 that inaugurated the first *Gründerzeit* as the major period of sociospatial transformation, in which Berlin was reimagined as the new imperial capital and as a major industrial metropolis (Hansen, 1995: 385; see Ribbe, 2002; Ritchie,

1999; Süss and Rylewski, 1999). Berlin's wholesale transformation into a modern city was all the more remarkable for a community that had remained a peripheral Prussian town well into the 19th century. While the city's status as the administrative capital of Prussia had spawned a neo-classical architectural renaissance in the early years of the 19th century, it was the abrogation of economic tariffs and the expansion of the German customs union in the 1830s which opened the city up to new economic markets and incipient industrialisation. As the textile, machine and metalworking industries gained prominence the city grew rapidly, doubling in population between 1849 and 1871. Even so, it was ultimately the proclamation of a unified German State in 1871 which was the catalyst for a new and accelerated phase of urban expansion that transformed the provincial capital into a true *Weltstadt*, an industrial and commercial centre of global significance. In the 30 years before World War I, the population of Greater Berlin doubled from two million to nearly four million. As the historian Peter Fritzsche points out, "...entirely new zones, grafted at sharp angles onto the old Prussian centre, indicated the city's rapid growth" (1996: 30). New factories spread west down the Spree River and, ultimately, in other directions as well. Around them clustered the largely proletarian precincts of Wedding in the north, Kreuzberg and Neukölln in the south, and a wide crescent of neighbourhoods extending north and east from Alexanderplatz. Industrial wealth also helped to create the middle-class suburbs to the west and southwest (Charlottenburg, Wilmersdorf, Schöneberg).

Whatever the case, it was overcrowding, poor housing and inadequate infrastructure that confirmed the city's status as a symbol of unfettered urban development. Statistically speaking, by 1871 over 162,000 residents in Berlin were living in one-room dwellings that, on average, housed over 7 people. In 1900, 43 percent of households lived in dwellings consisting of a single room, and 28 percent in two-room dwellings; in 1925, over 70,000 Berliners still lived in dilapidated cellars (Hegemann, 1979 [1930]: 337, 463). While, in 1910, the average number of residents per building was 8 in London, it was 5 in Philadelphia, 9 in Chicago, 20 in New York, 38 in Paris and 78 in Berlin (Hegemann, 1979 [1930]: 19; Eberstadt, 1920: 6). Many buildings were without basic sanitary arrangements and, in 1918, the Reich Statistical Office found that nearly a third of small dwellings lacked a kitchen and a toilet (McElligott, 2001: 70). In 1905, infant mortality in the working-class neighbourhood of Wedding remained a staggering 43 percent (Jelavich, 1993: 12).

The ultimate origins of Berlin's housing crisis were not, however, a product of the city's uneven and belated modernisation. If anything, they can be traced back, as Werner Hegemann's monumental political history of architecture and housing in Berlin argues, to the middle of the 18th century and the reign of King Frederick II (1740–1786) (see Hegemann, 1979 [1930]). According to Hegemann, this can be attributed to two factors in particular. First, the revision of Prussian mortgage laws in 1748 and 1750 which led to widespread speculation, rising rents and exorbitant property prices and which were, in turn, exacerbated by a series of major military campaigns in the 1750s and 1760s.[4] Second, the systematic neglect of Berlin's infrastructure which contributed to a growing shortage of housing. Whilst Frederick II devoted his own attention to a small number of major architectural projects, rent rises made the construction of multi-apartment houses attractive as the owners of armament factories invested their recent returns on new three- to four-storey buildings. These developments were nevertheless limited and it was the *militarisation of housing* during Frederich II's reign which established the basic conditions for Berlin's later reputation as a city of 'tenement barracks' (what would become known as *Mietskasernen*).

Frederick II took inspiration, in this context, from Paris where high-density accommodation for soldiers was common and he began a programme in Berlin which focused on the construction of military barracks. Smaller one- to two-storey homes were indiscriminately torn down across the city to make way for the new buildings. "In these barracks," writes Hegemann, "married soldiers were billeted with their families and it was, in this context, that the Berlin family was systematically drilled in a military way of life" (1979 [1930]: 114). The new housing thus served a wider *disciplinary* function in a city where, by the 1780s, every third resident was living in military accommodation. For the writer, geographer and historian Karl Friedrich von Klöden, who was himself born in a barrack in 1786, the punitive nature of the experience and the *martial urbanism* it endeavoured to produce was not lost on Berlin's inhabitants who often referred to the barracks as 'prisons' (*Zuchthäuser*) (Klöden, 1874). In practical terms, a number of disciplinary arrangements were central to life in the barracks. At night, the gates were locked in order to prevent desertion. In daytime, entry was strictly controlled and written permission through a '*visum*' was mandatory. Many of these 'regulations' were later adapted and applied to the large tenement blocks that housed many of Berlin's working-class residents in the late 19th and early 20th centuries.

In the end, it was the renting out of two barracks in the mid 1790s to local citizens in search of housing that first established the rental barrack (*Mietskaserne*) as an important feature of Berlin's built environment. Whilst it was, in fact, a later version of the Berlin tenement block that became intimately tied to successive rounds of creative destruction from the middle of the 19th century onwards, the intervening period was nonetheless marked by further speculation and growing inequality (see Vasudevan, 2011a; Geist and Kürvers, 1980–1989). As Johann Friedrich Geist and Klaus Kürvers have pointed out, in the period between 1815 and 1828 the number of affordable homes in Berlin decreased by over 50 percent. During the same period, the percentage of empty homes had increased from 0.75 to 4 percent (1980: 129). Increasingly precarious tenants responded, on the one hand, by adopting a range of informal practices that included the occupation of new buildings which were still under construction and where the mortar had not set, a practice that later became known as '*Trockenwohnen*'. On the other hand, they also displayed a growing willingness to challenge local authorities and resist planned evictions (Kowalczuk, 1992: 245, 236; see Geist and Kürvers, 1980).

The rapid development of the city in the second half of the 19th century represented a further victory for landlords and speculators over planners, social commentators, reformers and scientists. In a well-known history of the German working class, the German Marxist and co-founder of the Spartacist League, Otto Rühle, argued that this was largely the product of a political system which guaranteed that at least half of all city councillors in Berlin were registered landlords. For Rühle, in this way a majority political block was secured whose "interest in high rents and land prices, rent seeking and controlled expansion ran against the wider public interest". "It was overcrowding and growing demand that was responsible for their profits," Rühle continued, "and they were, unsurprisingly, the main focus of their housing policy. The result was the [Berlin] tenement house" (1930: 382). If there is much to recommend in this view, other commentators – including Hegemann himself – have singled out the Hobrecht plan of 1862 as a key moment in the transformation of Berlin in the 19th century (Bernet, 2004; Ladd, 1990; Strohmeyer, 2000). The extension plan for the city of Berlin – drawn up by the then inexperienced engineer James Hobrecht – focused on the circulation of traffic and future development outside the built-up core of the city. The final result was a vast and regularised grid of city blocks that were linked to existing roads and property lines. State officials attempted to regulate the acquisition of land where streets were planned but the plan did not provide any provision, as Brian Ladd (1990) has argued, for controlling "what was or not built on the privately owned land that the streets traversed" (81). A land speculation boom ensued as landowners sought to maximise windfall profits. The result was the typical Berlin *Mietskaserne*, a large block which was traditionally 5 storeys high and extended to the back of the lot, and was only broken up by a series of tiny courtyards that could be a small as 5.3m × 5.3m, the minimum necessary to comply with fire regulations.[5] Poor living conditions, disease and overcrowding were commonplace.

By the early 1860s, protest against housing scarcity and forced evictions had also become a regular occurrence in many districts of Berlin. As a new crisis in housing intensified in the immediate wake of the Hobrecht plan, landlords exploited the opportunity with many capping tenancy agreements at three months, others demanding that tenants pay a third of their annual rent in advance (a practice that was known as '*quartaliter pranumerando*'). Almost half (49.6 percent) of all renters in the city were thus forced to move as thousands of families poured onto city streets with all their possessions. Protests soon erupted and, in June 1863, the planned eviction of a Kreuzberg barkeeper led to a series of violent riots across the neighbourhood (see Lindenberger, 1984). What became known as the *Moritzplatz-Krawalle* began when a local barkeeper on Oranienstraße was issued with an eviction notice on the grounds that he had illegally installed an iron stove in his bar. In response, he placed a large placard in the window of his establishment that drew attention to the terms of his imminent eviction (Kowalczuk, 1992: 240). Crowds soon gathered in front of the bar and within hours an angry mob had attacked and vandalised the nearby home of the landlord. Riots broke out and it was only in the early hours of the morning that the police were able to take control of the local neighbourhood. Over the following days, further violent confrontations with the police took place as barricades were erected across Kreuzberg (Lindenberger, 1984: 46–48). The police responded by issuing a series of curfews and closing a number of city gates in order to prevent protesters from escaping. After four days of rioting, the police were finally able to quell the violence. In total, over 426 residents were arrested and dozens were seriously injured. The city was also forced to set up a special committee to oversee and settle disputes over rent while a conservative newspaper, *Die Neue preussische Kreuzzeitung*, warned of the rise of a "new form of social protest" in response to the "housing question" (see Lindenberger, 1984).[6]

There were, however, few immediate solutions to the 'housing question' which, if anything, only gained wider significance in the decades that followed. The victory of the Prussian military in the Franco-Prussian war and the founding of the German Reich in 1871, coupled with Hobrecht's earlier extension plan, led to a new crisis characterised by widespread speculation and predatory rent seeking. Of course, rent seeking, as David Harvey reminds us, is nothing more than a polite and rather neutral-sounding way of describing what he has often referred to as "accumulation by dispossession" (2014: 133). For many Berliners in the early 1870s, dispossession and displacement were everyday matters of fact as affordable homes disappeared in a frenzied cycle of panic buying and rising rents (see Nitsche, 1981; Geist and Kürvers, 1984: 121). Between 1871 and 1873, the return on rental income in Berlin rose by over 190 percent, while many residents were forced to pay rents that had doubled or tripled in the same period. At the same time, attempts to reabsorb capital surpluses found material form in a boom that transformed the numerous fields and meadows which surrounded the city into building plots and construction sites.

By April 1872, over 15,000 residents in Berlin were homeless (Glatzer, 1993: 80). Many responded to the housing crisis through a repertoire of informal makeshift strategies. This included the practice of *Trockenwohnen* where tenants moved in and out of brand-new buildings as the mortar set in exchange for nominal rents.[7] Others became "hot bedders *(Schlafleute)*" in which they paid rent for a place to sleep – often in shifts – in vastly overcrowded apartments. Yet others occupied and used vacant buildings (the first squatted buildings in Berlin were in April 1872 on Zionskirchestraße and Mohrenstraße), stables, empty carts, abandoned railway carriages and summer houses as temporary emergency shelters (Rada, 1991: 152). Some families even opted to camp out on newly-paved streets (Kowalczuk, 1992: 244). At the same time, a series of precarious squatter settlements were set up on the remaining open fields outside the gates of the city, most notably around the Halle and Kottbuser gates in the south and the Frankfurter and Landsberger gates to the east (Poling, 2014: 258). By March 1872, over 80 shacks had sprung up on the Rixdorfer Feld while the "*Freistadt Barackia*" that appeared outside the Kottbuser gate housed approximately 150 families (Bernstein, 1907: 230; Lange, 1976: 133).[8] There were a number of smaller encampments as well. In some quarters of the press, the squatter settlements were greeted in terms that drew on a racialised imaginary in which they were compared to a "Negro city in innermost Africa".[9] In contrast, more positive portrayals drew attention to the spirit of patriotism, self-reliance and independence that, in their eyes, characterised the new informal communities (see especially Ring, 1872).[10] If these celebratory accounts shared a number of features with writing about German settlers in far-flung colonial settings, it also became clear that many Berliners were increasingly willing to confront the very agents of their displacement through an engagement with a range of contentious political actions (see Poling, 2014: 267). Whilst such resistance to housing inequality constituted, on the one hand, a necessary response to immediate and pressing needs, it also represented, on the other hand, a new tentative opportunity to cultivate and develop other forms of association and solidarity (Kowalczuk, 1992: 245).

It was the eviction of a carpenter from an apartment on Blumenstraße in the district of Friedrichshain on 25 July 1872 that brought the tensions surrounding Berlin's housing crisis into sharp relief.[11] As the *Neuer Social-Demokrat,* the newspaper of the General German Workers' Association, reported, "for the next three days, Berlin was in an almost uninterrupted state of revolt".[12] Large crowds gathered across Friedrichshain and were soon attacking police with stones and other missiles. The protests lasted long into the evening and, while the next day began quietly, the eviction and destruction of a squatter settlement near the Frankfurter gate precipitated a new wave of resistance as barricades were erected in Friedrichshain and police were once again attacked by large crowds. As the protests stretched into a third day, local military units were put on standby. The homes of a number of landlords were offered police protection, which only exacerbated tensions and led to further violent confrontation between local residents and armed police units. It was only as the day came to an end that the police were able to contain the unrest as the protests ended with the same abruptness that they had begun.[13] Upwards of 150 protesters were wounded during the three days of rioting, whilst an additional 102 police officers were injured. Of the 100 protesters who were eventually arrested and charged, approximately 30 were sentenced to lengthy prison terms for breaching the peace (Kowalczuk, 1992: 247). The *Spenersche Zeitung* also reported that at least one *agent provocateur* working on behalf of the police was exposed during the 11-day court case (quoted in Nitsche, 1981: 55).

If the Blumenstraße riots represented a brief and violent response to the housing question, they did not lead to a sustained episode of contentious claim-making (see Tilly and Tarrow, 2006). The period after the riots was characterised, if anything, by further repression and deepening immiseration. The incoming Berlin police president, Guido von Madai, insisted in August 1872 that the remaining squatter settlements outside the city gates of Berlin were to be evacuated and destroyed by the middle of September. Any inhabitant who was unable to secure alternative shelter was to be rehoused in a workhouse for the homeless.[14] Despite representations from the squatters, the encampments were cleared and, in many cases, brutally dismantled by the police (Weipert, 2013: 39; Geist and Kürvers, 1984: 120). By the end of September, only a few shacks remained while hundreds, if not thousands, of Berliners were once again homeless, many of whom were unwilling to take shelter in workhouses. There were numerous suicides and, for a large number of city inhabitants, the situation only worsened as the traditional October 1 moving date approached. In Berlin, tenants typically moved at the beginning of April and October and, in 1872, over 53 percent of the city's renters were forced to find a new residence as city streets were once again full of families pushing carts with all of their possessions, an image that was later captured to great effect by the local artist Heinrich Zille.

The housing crisis of the early 1870s reached a peak in 1873 though the structural conditions that were responsible for the uneven geographical development of Berlin were largely unresolved. "Capital," writes Harvey, "represents itself in the form of a physical landscape created in its own image, created as use values to enhance the progressive accumulation of capital." Not only is surplus value generated, according to Harvey, through the production of housing but it is also pumped into the housing sector through a process he refers to as "capital switching" (1989: 83, 71). As Harvey argues, there exists both a 'primary' circuit of capital that encompasses the production of good and services and a 'secondary' circuit of capital marked by the production and reproduction of the built environment, including housing. For Harvey, the secondary circuit serves as an "overflow tank" into which surplus overaccumulated capital can be switched in the event of a crisis that prevents it from being profitably reinvested in the primary circuit. In this way, capital is effectively "parked" until such time as the conditions of overaccumulation have eased or there is an overinvestment in the built environment (see Aalbers and Christophers, 2014: 379; see also Harvey, 1982, 1985).

While the architecture of Harvey's argument may sound forbiddingly abstract, it is in housing, and the built form more generally, that these abstractions are most visible and most material. For many 19th-century Berliners, it was the social life of the tenement block (*Mietskaserne*) that, more than anything else, anatomised the iniquities of capitalist accumulation. Overcrowding and poor housing conditions remained a constant in the period that led up to the outbreak of the World War I. This was despite a growing political acknowledgement of the housing question in Berlin and elsewhere in Germany and an acceptance among the political elite that "poor housing was a concern of government". From the 1880s onwards, housing assumed a new prominence amongst planners and social reformers across Germany though the political strength of various economic interests often stymied reforms. In the case of Berlin, little real progress was made despite the efforts of physicians and public health reformers to rewrite the city's 1853 building code with a view to introducing new provisions on basement apartments, building height and the minimum size of courtyards. At the same time, the proportion of the city's budget devoted to housing remained a fraction of the money set aside in other major German cities (Ladd, 1990: 141, 155, 156). Nevertheless, riots, protests and other forms of contentious action related to housing were still episodic at best and often coincided with periods of heightened scarcity and insecurity. It was only World War I and its aftermath that led to the emergence of organised resistance to housing inequality and a more systematic approach to city planning (see Frisby, 2001; Scarpa, 1986).

Tenant Trouble: Rent Strikes in Weimar Berlin

The struggle over housing in Berlin in the 19th century – however fractured and uneven – demonstrated that the field of political action in the city was not confined to a narrow circumscribed terrain of activity separated from the strictures of everyday life, practised only at particular moments and by a limited segment of society. As a number of scholars of modern German history have argued, the circumstances under which ordinary people lived their lives, and the specific patterns of sociability and subjectivity that they generated, played an important role in the development of alternative political forms (Eley, 2002: xiii; see also Eley, 1989; Ludtke, 1993, 1995). The scholarly emergence of an anthropologically-oriented approach to the history of everyday life (*Alltagsgeschichte*) drew attention, in this way, to the material conditions of daily existence and the manifold terms in which these conditions were experienced, accommodated and contested. The first wave of this work focused, in particular, on the Kaisserreich (1871–1914) and the Weimar Republic (1919–1933) and the evolution of working-class agency and consciousness. Studies zoomed in on the everyday geographies of working-class lives and livelihoods, and the often informal politics surrounding work, food and housing became key sites of inquiry (Canning, 2002; Niethammer and Brüggermeier, 1976; von Saldern, 1984, 1995; Davis, 2000). As the historian Geoff Eley (2002) has argued, "working-class everydayness" represented the starting point for a more sophisticated appraisal of "the roughness and disordered transience of much working experience, with its dependence on informal economies, casualised labour markets, improvised domestic arrangements and crime" (xiv).

While traditional Marxist historiography dismissed the everyday as a scene of alienation and false consciousness, scholars such as Adelheid von Saldern (1995) and Belinda Davis (2000) have shown how ordinary struggles around food and housing carried their own political significance, dissolving boundaries between workplace, household and community. For Davis, it was food riots in Berlin during World War I which highlighted the cardinal significance of the city's poorer inhabitants in political decision making. According to Davis, it was the everyday experience of the food crises and the various forms they assumed that transformed the city into a veritable theatre of protest (2000: 5, 7). Von Saldern, in turn, demonstrated, in a series of groundbreaking studies (1984, 1993, 1995) that the 'housing question' of the late 19th and early 20th centuries was an undertheorised site of political action and that there were significant gaps and dissonances between official policies and reforms and the way in which they were encountered and experienced by those to whom they were addressed.

It is against this background that the historical geography of housing and precarity developed in this chapter must be situated. The everyday struggles of Berliners over housing were not, in other words, an epiphenomenon of wider economic structures but were, in fact, central to debates over the meaning and identity of Berlin as a modern metropolis. Whilst the 19th century bore witness to a range of protests over housing inequality, these struggles took on an even greater significance during the Weimar Republic, a period of modern German history typically described as the troubled interlude between two eras of greater historical significance: the Wilhelmine Kaiserreich, which created a unified nation, and the Third Reich, which destroyed it (Kaes et al., 1994: xvii). According to these accounts, 'Weimar' is frequently caricatured as a desperate and irredeemably compromised experiment in democracy, whose failings would end up having profound consequences not only for Germany but also for the world (see Gay, 2001; Laqueur, 1974). Writing, however, in response to these prescriptions, Peter Fritzsche has suggested that the Weimar Republic was neither a gamble nor a slapdash experiment, but rather a "laboratory of modernity" (1996: 631). For his part, Fritzsche was primarily interested in what were indisputably 'cultural' experiments and, at face value, his understanding of the Weimar era as a period of enormous creativity and experimentation in any number of fields, including architecture and housing, has become something of a historical commonplace (see Vasudevan, 2006, 2007). And yet, as much as Berlin represented, according to this view, the crucible of Weimar's 'experimental urbanism' and the "icon of German mass culture and modernity", widespread material improvements remained limited and uneven despite a number of reforms and the construction of new working-class housing, especially under the auspices of Martin Wagner who was chief city planner between 1926 and 1933 (Hake, 2008: 1; see Frisby, 2001; Pugh, 2014; Scarpa, 1986).

The economic recession of 1929 only led to renewed deterioration in housing conditions across Berlin and the emergence of an organised campaign of rent strikes and occupations in the early 1930s. Despite the scale and intensity of the strikes, they have remained an understudied chapter in the history of housing activism in Germany. As Uwe Rada (1991) has argued, "while there are hundreds of metres of shelves filled with treatises on housing inequality, social reform, architectural history and urban planning, a mere 20cm have been devoted to the victories and defeats of tenants in their fight against slum landlords and speculators" (169). It was ultimately housing activists in Berlin – from local tenant groups to squatters – that, beginning in the late 1970s, drew attention to the history of rent strikes in Weimar Berlin and demonstrated the significance of housing as a key site of political action (see Kowalczuk, 1992; Nitsche, 1981). As their research painstakingly showed, roughly 3000 apartment blocks in Berlin – and perhaps many more –were involved in rent strikes in 1932 and 1933, while hundreds of thousands tenants participated in earlier initiatives in the 1920s (Rada, 1991: 169).

It was the collapse of the German Reich in the final months of World War I that was ultimately responsible for the emergence of a coordinated movement for tenants rights across Germany. Whilest a detailed exegesis of the events that led to the abdication of Kaiser Wilhelm II and the proclamation of a new German Republic is beyond the compass of this book, it was the mutiny of sailors in Kiel and Wilhelmshaven in early November 1918 that triggered the rapid disintegration of the Reich. For a brief period, political power was effectively in the hands of the sailors', workers' and soldiers' councils that sprang up and occupied barracks and factories across Germany. A Socialist Republic was declared in Berlin on 9 November, while the new councils seized political control in a number of cities across Germany, including Leipzig, Bremen and Hamburg. The council movement was unable, however, to maintain its power for long and, on 16 December 1918, the conference of Workers' and Soldiers' Councils in Berlin voted to set up elections for a parliamentary National Assembly which effectively divested councils of any real power. Dissatisfaction amongst many workers with the course of the 'revolution' precipitated a new wave of unrest in early January 1919 in what became known as the Spartacist uprising. The uprising itself was swiftly crushed by the leadership of the Social Democratic Party of Germany (SPD) who had refused to work with the revolutionaries. The SPD ordered the use of anti-republican proto-fascist militia known as the Freikorps who suppressed the rebellion and, on 15 January 1919, brutally murdered its leaders, Rosa Luxemburg and Karl Liebknecht. A few days later, a constituent National Assembly was elected (the SPD was the largest party) and, in February 1919, a new Republic was officially recognised in the small provincial town of Weimar (Ryder, 1967; Weipert, 2013; see Kuhn, 2012).

For many Berliners, the reality of housing in the early years of the Weimar Republic did not keep pace with the social programme of the ruling SPD party which emphasised, in theory at least, material improvements, social reforms and large-scale urban planning. In practice, however, existing patterns of uneven geographical development were largely untouched by the war as landlords exploited a new crisis in housing through rent rises and further speculation (Rada, 1991: 169–170). Tenants in Berlin had little choice but to form new councils and organisations, especially in the wake of the collapse of the workers' and soldiers' councils which had attempted to introduce rent controls and bring housing under common ownership. While earlier efforts were made to regulate the housing market during the war and, in so doing, contain a potential source of social unrest, these measures were largely ineffectual and tenants remained, for the most part, at the mercy of landlords.

It is hardly surprising, therefore, that tenant organisations gained widespread popularity across Berlin and that local neighbourhood committees and councils developed a repertoire of contention which challenged the authority of landlords. Tenants campaigned, in particular, for deductions in rents to cover basic repairs whilst eviction resistance networks were established in order to protect residents against arbitrary expulsions. *Die Rote Fahne*, the main organ of the Spartacist League and the Communist Party of Germany, was already reporting by the end of 1918 that "an appeal had gone out to tenants living in the northern suburbs to resist any increases in their rent and to organise against evictions".[15] A few days later, the same newspaper reported the declaration of a rent strike in the district of Weissensee, one of many spontaneous actions that erupted in the immediate post-war period.[16] The turn to contentious politics by the new tenant councils generated tensions between activists who favoured direct action and other more moderate organisations such as the Greater Berlin Tenant's Association (*Gross-Berliner Mieter-Verbandes*) who continued to negotiate with local authorities. These tensions were, in the end, responsible for the splitting of the movement as a radical grouping of tenant assemblies and councils chose to continue in a more direct line of resistance (Rada, 1991: 172).

It was new legislation on the regulation of rents that proved decisive in radicalising the tenants' movement across Berlin. What became known as the 'Reich Tenants' Law' (*Reichsmietengesetz*) was proposed as an expansion of existing provisions that had been put in place during the war under which rents were capped. By the end of 1920, new draft legislation would have made it possible for rent increases upwards of 200 percent. Tenants responded with a wave of protests and, on 1 April 1921, thousands in the city and elsewhere in Germany began a rent strike.[17] The strikers demanded control over the administration of rents, the seizure of vacant properties and the immediate restoration of homes in need of repair (Hermann, 1925). While local actions yielded mixed results, the strikers were successful in introducing rent controls which were passed into law in April 1922 and which, in the case of Berlin, existed in a revised form until 1988. The tenant movement's immediate success was short-lived, however, as the onset of hyperflation by the end of 1922 neutralised the effectiveness of the new law. What did survive, however, was an action repertoire that produced local geographies of engagement and solidarity embedded within city neighbourhoods that would prove useful as a new wave of protest descended on the city in the early 1930s. As in the case of wartime food riots, it was working-class women who played a central role in the radicalisation of the housing movement and whose identification as political agents was dependent on an understanding of the city as a site of protest and revolt and their own actions as insurgent forms of citizenship which treated housing as a fundamental right (Davis, 2000: 3; see Holston, 2008).

Protests around housing insecurity took on an even greater significance and intensity in the early 1930s as the economic collapse of 1929 triggered widespread unemployment. Decreasing wages and new taxes on basic consumer goods in turn exacerbated a cost-of-living crisis that left many Berliners facing eviction (Nitsche, 1981: 160). Tenants responded through a new direct action campaign that was launched at an assembly on Swinemünder Straße, only a few doors down from one of the first squatted houses in Berlin originally occupied during an earlier wave of protests in 1872. On 27 July 1932, 14 houses on Swinemünder Straße, representing over 300 families, agreed to go on a rent strike at the beginning of August. They were fighting for a 30 percent reduction in their rents and were soon joined by others houses in the districts of Mitte and Prenzlauer Berg (Rada, 1991: 174–175). Tenants living in a former jail on the Molkenmarkt began their own strike on 1 September. Over 110 families lived in the 'bed bug castle' ('*Wanzenburg*') in cells that were as small as 2m × 4.5m. The tenants demanded a 50 percent reduction in their rents and the immediate renovation of the building. The landlord responded by issuing eviction notices and a bailiff was ordered to expel tenants on the morning of 14 September. A large crowd gathered in the courtyard of the building and prevented the bailiff from carrying out the evictions.[18] Despite a campaign of intimidation by the landlord, they eventually agreed to subsidise a programme of renovations and to lower rents by 40–42 per cent. Any outstanding rent payments were waived and the strike ended less than a month after it began.

The repertoire deployed during the strike only emboldened housing activists across Berlin who came together at a large conference at the end of September 1932 attended by over 1000 delegates (Rada, 1992: 177). At the same time, new strikes were declared in a number of houses including one on Köpenicker Straße 34/35 which soon became a key symbol of the wider movement. Of the block's 63 households, 58 joined the strike adopting similar demands to their comrades in the *Wanzenburg*. *Die Rote Fahne* reported that the house's 160 residents were limited to the use of six working toilets while a number of children suffered from acute tuberculosis. The local tenants' assembly set up a stall in the courtyard where they 'scientifically' displayed the various cockroaches, beetles and bedbugs that they had collected from the house.[19] The tenants were successful in securing an amnesty for outstanding rents, a withdrawal of planned evictions and a commitment from the landlord to undertake much-needed renovations. They were unable, however, to reduce their rent by more than 10 percent and remained determined to continue striking. Neighbourhood demonstrations and other initiatives were organised as the citywide movement expanded. New tactics were devised in response to forced evictions and growing police violence. A popular tactic involved re-occupying apartments in the immediate aftermath of an eviction once the bailiffs and police units had left, as landlords were legally required to submit an application in order to carry out a new eviction and this could take at least eight weeks to process (Kowalczuk, 1992: 251–252). Other evicted tenants set up large squatter encampments on the outskirts of the city in garden allotments, a subject that was a feature of Bertolt Brecht's 1931 film, *Kuhle Wampe*. By the early 1930s, makeshift shacks and huts on the city's allotments were already housing roughly 120,000 of Berlin's residents or about 2.8 percent of the city's population (Urban, 2013: 221).[20]

On 29 December 1932, the tenants at Ackerstraße 132, Berlin's largest tenement block, voted almost unanimously to go on strike. The final outcome of the strike nevertheless remains unclear as the *Die Rote Fahne* had been banned in the aftermath of the February 1933 Reichstag fire. As a number of historians have argued, despite the efforts of the *Die Rote Fahne* to record and document the rise of housing activism in Berlin, it has remained a challenge to reconstruct the full history of rent strikes in the final years of the Weimar Republic (see Kowalczuk, 1992). There is plenty of evidence to support the view that the strikes took in thousands of apartment blocks across the city. In the neighbourhood surrounding the Stettiner Bahnhof (now know as Nordbahnhof), over 312 blocks and 14,615 tenants were known to be on strike in October 1932 and it is entirely plausible that, across the city, over 3000 houses were actively striking by late 1932.[21] The tenant movement came to an abrupt end, however, as the seizure of power by the Nazis in the wake of elections in March 1933 led to a violent crackdown on left-wing activists, many of whom had been active in rent strikes and eviction resistance networks alongside other alternative youth groups (Weipert, 2013).[22] While radical forms of political action were crushed by the Nazis, other aspects of the 'housing question' were co-opted by a corporatist state that envisaged a *völkisch* national community and the total forcible coordination (*Gleichschaltung*) of Germany society. Moderate housing organisations such as the Confederation of German Tenant Groups (*Bund deutscher Mietervereine*) were, in this way, re-engineered and brought into line with Nazi policy and, it was only many years after World War II that an alternative politics of housing would re-emerge to find a new voice in a divided Germany.

Re-building a Divided City

When the Nazis seized power in 1933, Berlin's reputation as "the capital of modernity" was already well established (Pugh, 2014: 26). For many Nazis, however, the city was synonymous with the Weimar Republic and it was, as the historian Andrew Lees has argued, "entirely plausible for [its opponents] to conflate the beginning of their dictatorship with a conquest of a corrupt and chaotic capital" (Lees, 1985: 170–4). Unsurprisingly, Albert Speer, Hitler's chief architect, was tasked with producing a new urban plan for a city that was to be the centrepiece of Hitler's German Empire. The plan was unveiled in January 1938 and was designed to 'spatially cleanse' the city of those populations and buildings that did not fit in with a National Socialist worldview. Speer's plan called for the construction of vast new structures situated on a classically symmetrical plan organised around two axial boulevards. While construction on the scheme to re-imagine Berlin as a monumental tribute to Nazi power began in 1939, the plan was largely thwarted by the outbreak of World War II though it nevertheless cast a "long shadow over plans for Berlin's post-war reconstruction" (Pugh, 2014: 27).

At the same time, the wholesale destruction of Berlin by the end of World War II was widely seen by planners as an opportunity to finish what the bombs began, "a one-time chance to finally overcome the *Mietskasernenstadt*" (von Beyme, 1987: 85). The war had, after all, reduced the city's housing stock by over 50 percent, leaving over 80 million cubic metres of rubble on city streets, while the stigma surrounding crowded apartment blocks had only intensified during the war as inner courtyards quickly became death traps in times of aerial bombardment (Bernet, 2004: 413). As planners, architects and politicians welcomed the further demolition of Berlin's *Mietskasernen* as a much-needed solution to the city's housing needs, they also sought to justify their actions as a means to distance not only themselves but also a 'new' Germany from 12 years of National Socialist rule (see MacDougall, 2011a).

The reality on the ground was not marked, however, by a decisive break with the past and was further complicated by the official division of the city and the creation of two German states in 1949. While preliminary plans for a 'new' Berlin were developed as early as 1946 by the modernist architect and former proponent of Weimar modernism, Hans Scharoun, these were rejected as too utopian in favour of a new plan devised by Karl Bonatz, another veteran of the Weimar era. The plan was never put into practice and, as the historian Carla MacDougall has argued, the failure of these plans was not simply a product of emerging Cold War tensions. They also failed because they ignored the everyday realities of Berliners who faced an acute shortage of housing. The plans called for the wholesale demolition of buildings when the immediate reconstruction of salvageable housing stock seemed, to many Berliners, a more suitable solution (MacDougall, 2011a: 54).

Plans to demolish Berlin's *Mietskasernen* nonetheless played an equally important role in the wake of the city's geographical partitioning. Governmental and ideological divisions became, on the one hand, "increasingly entrenched over the course of the 1940s and 1950s, as architects and planners were obliged to situate themselves physically and aesthetically along Cold War divides" (Pugh, 2014: 27). According to this view, a divided Berlin assumed a new ideological role as a stage for competing visions of urban planning and housing (see Castillo, 2005, 2010). On the other hand, early showcase developments such as Stalinallee in the East and the Hansa Viertel in the West were contingent on the further destruction of an existing city and, along with it, thousands of homes that could have provided much-needed housing. In both cases, it was architecture that became the medium through which the promise of a new political community was imagined (see Koshar, 2000: 11). The construction of Stalinallee in the early 1950s, a major boulevard that extended eastwards from Alexanderplatz lined with 7- to 10-storey apartment blocks, was thus predicated on establishing a continuity between a socialist future and an earlier German style of architecture. The apartment blocks featured designs that referenced works by native Berlin architects, most notably Karl Friedrich Schinkel, while the wider development, according to East Berlin's urban plan, was linked to the city's historic Unter den Linden boulevard. In this way, as one historian has recently argued, the governing SED was able to connect the GDR "literally and symbolically with the historic traditions of Germany in its national capital, thereby symbolically making the claim that it, and not the Federal Republic, was the more legitimate Germany" (Pugh, 2014: 38).

The reconstruction of West Berlin was also framed in explicitly political terms with the development of the Hansa Viertel in 1957 representing an opportunity to showcase a viable modern alternative to Stalinallee. Whereas the latter was based on forging an historic antifascist aesthetic, the former was conceived as a celebration of West Germany's postwar prosperity. The reconstruction of the Hansa Viertel, a badly damaged district of the West on the edge of the Tiergarten, was initiated in 1951 as part of the International Building Exibition or *Interbau*. While the project was ostensibly devised to produce a new residential district, it ultimately served an exhibitionary function showcasing the West's embrace of a modern consumer economy and a liberal political framework (Stephens, 2007: 2–3; see Moeller, 1997; Pugh, 2014). The exhibition thus placed particular emphasis on the construction of architectural 'models' that, according to its officially-stated objectives, highlighted "the technology and design sophistication of the free world in the diversity of its forms" (quoted in Pugh, 2014: 51). The ground plan for the Hansa Viertel was designed by two Berlin architects, Gerhard Jobst and Willy Kreuer, who in turn established an expert panel which was responsible for inviting a group of 53 architects to design and complete projects for the exhibition. The list included a number of prominent modernist architects such as Aalvar Aalto, Walter Gropius, Oscar Niemeyer and Max Taut. The overall plan called for the construction of 1300 units that adhered to well-established modernist principles. There were large amounts of open green space while the individually-designed buildings – from single-dwelling houses to large high-rises – were deliberately dispersed to create a low-density district. Any reference to the neighbourhood's earlier history and layout were conspicuously avoided (Pugh, 2014: 51).

The *Interbau* opened in 1957 and received over one million visitors in just three months. The architectural show was accompanied by a special exhibit, "The City of the Future" (*"Die Stadt von morgen"*), which provided a detailed model for the reconstruction and renewal of West German cities (MacDougall, 2011a). While the exhibit focused on the relationship between postwar urban development and the establishment of 'real democracy' in Germany, it also depended on a familiar *moral imaginary* in which the ills of the contemporary city were relocated to the 19th-century *Mietskasernenstadt* and its predominantly working-class inhabitants.

At stake here was the very health of the modern city dweller for whom the over-crowded Berlin tenement represented "a creeping poison" (Ottersky, 1957: 432). The hygienicist echoes of earlier anti-urbanist critiques are hard to miss here as the *Interbau* envisaged the idealised moral rehabilitation of West German society through the "physical form of housing" (MacDougall, 2011a: 66). To do so, how-ever, depended on the creative destruction of a working-class city and the whole-sale dismantling of its everyday geographies and their enduring social meaning. Dozens of older tenement blocks were therefore razed to the ground to make way for the *Interbau* whilst over 700 residents and a number of local businesses were displaced (Bodenschatz, 1985: 70).

The *Interbau* provided a veritable blueprint for the spatial transformation of Berlin in the postwar period that elevated a modern vision of clean healthy living at the expense of an older, irrevocably damaged version of the city. Scholars have tended to focus, in this context, on the propagandistic aspects of the exhibition and its relationship to a predominantly Western discourse on architectural mod-ernism which helped to establish the legitimacy of West Berlin as a 'western-oriented city'. "Photographs of the rebuilt Hansa quarter," writes Emily Pugh, "communicated both a powerful message of renewal and the abandonment of the forms and ideals of the past in favour of a promising and progressive future" (2014: 58; see Castillo, 2005, 2010). Less attention has been paid, however, to the iniquitous social geographies that underwrote the exhibition and the large-scale projects that followed, many of which remained either unaffordable to the city's working-class inhabitants or resulted in their displacement. As early as 1956, local authorities in the district of Kreuzberg had produced an elegant booklet entitled "*Wir bauen die neue Stadt*" ("We are building a new city") which documented the proposed reconstruction of the Luisenstadt district as an oppor-tunity for the "organic and functional renewal of the area's urban fabric. In this way, many errors of the past may be overcome" (1956: 8). The booklet singled out an earlier 19th-century caricature of Berlin as a source of overcrowding and poor living conditions and said that "new sound housing developments" would be constructed which placed particular emphasis on the separation of city living according to "function and character" (1956: 7). A number of additional improve-ments (green spaces, vocational schools and playgrounds) were also identified, though for whom the development was ultimately conceived remains less clear.

In this way, the *Interbau* served as a catalyst for the urban regeneration of West Berlin from the late 1950s onwards. The city adopted an official renewal policy that favoured the further demolition of older *Mietskasernen* and an emphasis on the construction of new housing. The history of this process can be divided into two main phases. The first phase focused on the building of large-scale housing estates on the outskirts of the city offering cheap rents through direct state sub-sidy. These new developments also targeted Berlin's remaining self-built settle-ments that continued to house – in many cases illegally – thousands of residents on the city's allotments.[23] As Florian Urban (2013) has recently argued, "modern

Berlin was literally built on the memories of informal neighbourhoods" (241). According to Urban, many were cleared to make way for modernist tower blocks including the Märkisches Viertel in Reinickendorf, the Falkenhagener Feld in Spandau (begun in 1962) and the Gropiusstadt in Neukölln (begun in 1960). The same also applied to other housing developments, including the Georg-Ramin-Siedlung in Spandau (begun 1957), the Paul-Hertz-Siedlung in Charlottenburg (begun 1961), and the Charlottenburg-Nord development (begun 1962) (see Urban, 2013: 241). While some allotment dwellers protested against formalisation and forced eviction, their actions were largely unsuccessful. Others left without major resistance and accepted the indemnity payments offered to them by local authorities.

The second phase emerged out of the economic recession of the 1960s which quickly brought an end to the building of massive modernist satellite cities. High rent costs and expensive financing prompted a *re-urbanisation* of capital and a pre-occupation with the renewal of inner-city neighbourhoods including Kreuzberg, Wedding, Neukölln and Schöneberg where there remained a high concentration of older housing stock. Whilst a preliminary 1961 report by the Ministry of Building and Housing identified over 400,000 tenement units across the city that were in need of repair, the city's 1963 official plan for urban renewal prioritised 56,000 units in total, 10,000 of which were deemed renewable, with the remainder slated for demolition (see Schwedler, 1964).[24] It was in this way that public housing developments were "transplanted into previously multipurpose *Gründerzeit* districts 'replacing' those historic districts with monofunctional modern districts" (Heyden and Schaber, 2008: 140). This policy, known as 'clear-cut or area reno-vation' ('*Kahlschlag- oder Flächensanierung*') involved the demolition of entire blocks of older low-rent housing in favour of new construction which ultimately served the interests of developers, private investors and city-owned housing associations (see Heyden and Schaber, 2008).

If the accumulation of capital, as David Harvey reminds us, "builds a physical landscape suitable to its own condition", it often does so, he adds, "only to have to destroy it" (1989: 83). In the case of Berlin's older tenement blocks, their wide-spread destruction in the 1960s and 1970s was a necessary condition for the building of an urban landscape that was favourable to "capital's further expan-sion and qualitative transformation" (Harvey, 2014: 155). It was, after all, more advantageous for the agents of urban renewal to destroy older housing and build new modern developments than it was to undertake much-needed renovations. Where housing was not destroyed in West Berlin, it was often intentionally neglected by its owners with a view to future demolition as new construction guaranteed heavy public subsidies. This was most notably the case in the neigh-bourhood of Kreuzberg where many of the city's vulnerable tenants struggled to secure decent affordable housing and were forced to either move to other badly maintained but affordable tenement blocks or face 'decanting' to one of the large satellite cities on the outskirts of Berlin. As a result, buildings that could have been rehabilitated and renovated in 1963 were, by the early 1970s, beyond repair while rampant speculation and local corruption contributed to an ever-intensifying crisis for residents dependent on modest to low-income housing (Figure 2.2). The uneven geographical transformation of West Berlin after the war was not without its critics. In 1965, the journalist and architectural historian, Wolf Jobst Siedler, published a vituperative attack on the modernist redevelopment of Berlin entitled *Die gemordete Stadt* (*The Murdered City*) in which a faceless modern cityscape was juxtaposed with an older historical version of Berlin which, according to Siedler, it was responsible for destroying. Siedler's criticism was, in the end, largely impressionistic and eschewed any analysis of the complex forms of sociability that shaped Berlin's working-class tenement blocks. A number of other texts nevertheless began to appear in the 1960s in West Germany that drew in a range of professionals in the social sciences, including the psychologist Alexander Mitscherlich whose 1965 book, *Die Unwirtlichkeit unser Städte: Anstiftung zum Unfrieden* (*Our Inhospitable Cities: An Incitement to Unrest*) launched a further critique of postwar architecture as "weak-spirited" and unable to generate a wider sense of place and community. Mitscherlich's student and former Adorno protégé, Heide Berndt, extended his project, and in a series of texts began to develop a critical urbanism in the 1960s that combined the psychoanalytical with the sociological whilst

Figure 2.2   The creative destruction of the Berlin tenement. Kreuzberg in the early 1980s. Skalitzer Straße between Kottbusser Tor and Görlitzer Bahnhof (Manfred Kraft/Umbruch Archiv).

acknowledging the influence of prominent American scholars including Jane Jacobs and Kevin Lynch (Berndt, 1968). While Berndt's work has received little attention outside of Germany (for an exception, see Klemek, 2011), it anticipated and contributed to the vibrant architectural culture that emerged in Europe in the late 1960s and early 1970s and pointed, in this way, to the articulation and exploration of a radically different politics of habitation (see Lefebvre, 2014).

Berndt's involvement in the student protests and anti-authoritarian revolt of the late 1960s also highlighted a growing radicalisation of professional planners and architects in Berlin and elsewhere in West Germany. While many professionals focused their criticisms on the need for greater citizen participation in the planning process, the emergence of new social movements increasingly connected the planning community with students, workers, local citizen initiatives and tenant groups. For many, the 'housing question' was no longer limited to a critique of modernist urban renewal but, as the remaining chapters of this book argue, was predicated on resisting the logics of capitalist accumulation and the official course of urban development upon which it depended. It in this context that a new generation of activists, architects and planners turned to the long and uneven history of housing inequality in Berlin as a means of identifying and reclaiming an alternative right to a different city while, in many cases, positioning their own actions – sometimes critically, sometimes less so – within and alongside an older repertoire of contention. In their eyes, the city and the built form, in particular, offered openings onto the production of "new urban sensibilities and collaborations" (Simone, 2005: 517). These openings also pointed, in the end, to a growing preservationist impulse that prompted many professionals to re-examine their understanding of Berlin's older housing stock and the wider patterns of work and sociability associated with its development.

Re-animating the 'Housing Question'

It was the Hamburg-born art critic Karl Scheffler, editor of the leading art journal *Kunst und Künstler*, who in 1910 famously described Berlin as "the capital of all modern ugliness" and as a "nowhere city that was always becoming and never being" (1910: 267). While the observation has, in recent years, become something of a commonplace in describing Berlin's complex relationship to its past, Scheffler's own target was the faceless city of grey tenements that, in his view, symbolised Berlin's rapid and uneven modernisation. The recent reappropriation of Scheffler has, in this way, tended to occlude the very forces – the political-economic reconfigurations – that were largely responsible for the city's rapid growth in the late 19th and early 20th centuries. At the heart of this chapter is an attempt to re-examine and re-construct the socio-economic dynamics underpinning Scheffler's opprobrious description of Berlin. This is, as the chapter argues, a history and geography of urban development and capitalist accumulation closely allied to repeated crises of housing in the city. At the same time, and for many Berliners, it was also widely understood and experienced as a source of political contestation and resistance that opened up spaces for change in the city, however fragile and precarious (Aalbers and Christophers, 2014; see Harvey, 2014). At stake here, as this chapter has shown, is a long history of oppositional practices adopted by Berlin's residents in response to housing insecurity, and it is in these very strategies that we find the contours of an alternative understanding of the city as a site of action and revolt.

The main aim of the chapter has been to retrace this history of dispossession and displacement, resistance and re-appropriation and to provide a supporting framework for understanding the necessary conditions that contributed to the emergence of Berlin's squatting scene and the various practices of resistance that they adopted. It is not my intention, however, to suggest that squatters were the natural successors to earlier rounds of protest and resistance to housing inequality in Berlin, nor do I wish to imply that there was anything inevitable about the spatial practices they adopted and re-imagined. Rather, if anything, I seek to relocate these practices within an expansive field of possibilities and potentialities and the fractured history of housing to which they are undoubtedly connected. This is, in many respects, a plea for greater historicisation and I am indebted – unsurprisingly perhaps – to a 1930 review of Werner Hegemann's history of the Berlin tenement by Walter Benjamin. For Benjamin, himself a longtime resident of Berlin, Hegemann's forensic critique of housing in Berlin was laudable though it betrayed, he believed, the limitations of a purely negative mode of analysis and critique. It has no eye, writes Benjamin, for the "indestructibility of the highest life in all things" and for "beauty even where it is most deformed" (Benjamin, 1972–1991, *GS* 3: 265). As Benjamin continues in a remarkable passage worth quoting in full,

This Last Judgement in the trial of the city of Berlin leaves one thing to be desired: ventilation... Even on the Day of Judgement, the fact that it happened so very long ago ought to count as an extenuating circumstance. For the passage of time has a moral dimension. This does not, however, lie in its progression from today to tomorrow but in the reversion (*Umschlag*) of today into yesterday. Chronos, like Leporello, holds in his hand a picture book, in which, one of the other, the days fall back into the past, thereby revealing their reverse side, their hidden unconscious life. This is the province of the historian. Goethe's theme is applicable here: '*Es sei, wie es wolle, es war doch so schön.*' It reconciles us with the past... Hegemann would not be the Jacobin he is if he let the genius of history lead him by the hand and show him the way to such mercifully physiognomic existence... It never occurs to him that these barrack-like tenements (*Mietskaserne*), slum housing though they are, have created streets whose windows reflect not merely suffering and crime but also the uniquely sad grandeur of the morning and evening sun, and that the city child has always extracted, from the stairwell and the street, substances as inalienable as those that the peasant boy finds in the stable and the field [...] (Benjamin, 1972–1991, *GS* 3: 265).

This is, of course, a dense passage that merits further attention in its own right as an example of Benjamin's idiosyncratic approach to historical writing. There is also a danger here that Benjamin's own reading of Berlin's urban landscape runs the risk of romanticising the city's tenement housing. As Benjamin himself made clear in an earlier radio broadcast in April 1930 on the history of housing in Berlin, entitled *"Das Mietskaserne"*, the rental barracks were "on the way out; through the abolition of the sombre and monumental stone building that has stood still, immovable, and unchanged for centuries". In its place, Benjamin argues, were new buildings made of concrete, steel and glass that will inspire people to be "freer, less anxious, and also less belligerent". And yet, what Benjamin described as the "struggle for liberation from the old, fortress-like, and gloomy barrack city" would ultimately end up returning to that 'older city' with a view to *reclaiming* it (Benjamin, 2014 [1930]: 61, 62).[25] It was, after all, squatters and other housing activists in the 1970s whose actions ultimately re-imagined and extended Benjamin's alternative urban 'physiognomy' by connecting the material 'substance' of the *Mietskaserne* and an older sedimented history of dispossession and displacement to a radically new understanding of the city as a space of possibility. As one squatter concluded, many years after Benjamin's original review, "Kreuzberg is not idyllic, the *Hinterhof* not romantic ... but love from yesterday is tucked away in cracks, and there are remnants here of opportunity and good fortune" (Ziegs quoted in Nitsche, 1981: 233). If nothing else, it was an awareness and understanding of these "remnants" that ultimately played a central role in the evolution of Berlin's squatter movement for whom they came to represent a source of protest, re-appropriation and experimentation.

Notes

1   *Gründerzeit* refers to the foundational period in German history after the unification of the country in 1870. It is often used to index a particular building style and the building boom which Berlin witnessed after unification.
2   Ring's article was one of a series of pieces that appeared in the pages of national publications which also included important illustrated family journals such as *Die Illustrierte Zeitung* and *Über Land und Meer*.
3   Construction of a customs wall around Berlin was begun in the 1730s. The original stockade was complete in 1737 and included 14 city gates. As the city expanded, the stockades and gates were moved. By 1801, the original wooden structure had been replaced by a 4-metre stone wall. A number of the gates were rebuilt – most notably the Brandenburg gate – and a number of new gates were added. It is outside these city gates that squatter settlements sprang up in the early 1870s though the wall itself was torn down in the mid-1860s. In fact, local inhabitants had already begun to dismantle the wall and create their own informal openings after the King ordered the wall's gradual removal in June 1865 (see *Die Neue preussische Kreuzzeitung*, 26.7.1865, 3; *Norddeutsche Allgemeine Zeitung*, 16.7.1865, 3; see also wider discussion in Poling, 2014).

4   The new laws established a ranking for the seniority of debt. Secured debt was given the first lien. Overall, the legal changes improved creditor rights and facilitated the verification of collateral. The end of the Seven Year's War in 1763 led to a collapse in speculative tradition and a credit crunch across northern Europe.

5   The minimum size requirements of courtyards corresponded to the turning radius of the fire department's wagons (see Ladd, 1990: 81).

6   *Die Neue preussische Kreuzzeitung*, 3.7.1863.

7   The predicament of Berlin's *Trockenwohner* at the turn of the 20th century featured prominently in Hans Fallada's posthumous novel *Ein Mann will nach oben* (1953).

8   The same story ran in a number of local newspapers, see *Spenersche Zeitung*, 17.5.1872; *Vossische Zeitung*, 18.5.1872, Beilage; *Neye preussische Kreuzzeitung*, 22.5.1872; also see *Deutsche Allgemeine Zeitung*, 14.7.1872, Beilage.

9   *Deutsche Allgemeine Zeitung*, 27.6.1872, 1465.

10 As Kristin Poling (2014) has recently argued, contemporary accounts of Berlin's informal settlements also echoed popular press articles from the 1860s that focused on the experiences of German settlers overseas. Berlin's shackdwellers were, in this way, re-imagined as "industrious homemakers turning a frontier landscape to productive use" (267). These accounts drew, in particular, on Germans living in the Americas although Poling argues that they also anticipated later colonial imaginaries that were deployed in the context of German settlement in southwest Africa.

11 According to the *Neuer Social-Demokrat*, it was widely reported that the eviction of the tenant, a Herr Harstark, was based on outstanding rent payments to the landlord. His three-year contract expired on 1 October 1872 and he had regularly paid his rent *quartaliter pranumerando*. The real reason for the eviction, it would seem, was that the fact that Harstark was subletting part of the apartment (see Nitsche, 1981: 46).

12 *Neuer Social-Demokrat*, 1.8.1872.

13 *Neuer Social-Demokrat*, 31.7.1872 and 2.8.1872.

14 *Spenersche Zeitung*, 24.8.1872; *Vossische Zeitung*, 8.28.1872.

15 *Die Rote Fahne*, 28.12.1918

16 *Die Rote Fahne*, 31.12.1918.

17 While the strikers boasted that over 250,000 Berliners were involved in the action, many scholars believe that a nationwide estimate of 300,000 strikers is a more plausible figure (see Rada, 1991: 174).

18 *Vossische Zeitung*, 14.9.1932.

19 *Die Rote Fahne*, 1.10.1932.

20 Figure quoted from *Lokalanzeiger*, 12.08.1933. As Florian Urban has recently argued, it is now clear that during the interwar period, the "vast majority of Berlin's approximately 103,000 garden plots were permanently inhabited" (2013: 230). While the overwhelming majority of these inhabitants were supporters of the Social Democrats (SPD) and the Communists (KPD), by the early 1930s there was also a small number who were members of the Nazi Party. During the 1930s and 1940s, the Nazis attempted to further regulate the settlements though, in the face of a serious housing crisis, they ultimately tolerated their existence. During this period, the garden plots also provided shelter for some of the city's persecuted residents including communists and Jews. As Urban notes, "the entertainer Hans Rosenthal, the actor Michael Degen, and the author Inge Deutschkron were among the most prominent Jews who survived

hidden in Berlin allotment colonies" (Urban, 2013: 248).The plots were also used as bases for resistance networks during the Second World War.

21   *Die Rote Fahne,* 30.10.1932.

22   There were, however, factions in the NSDAP that did offer a modicum of support to the rent strikes especially in Berlin though these initiatives ended as they seized power in 1933.

23   Unlike their counterparts in the West, East Berlin authorities tended to take a pragmatic approach and avoid large-scale evictions though tower block housing estates were constructed on allotments and similar compensation schemes were used to pacify their former tenants (see Urban, 2013).

24   By the end of the war over a third of West Berlin's apartment units had been destroyed (Der Senator für Bau- und Wohnungswesen, 1957: n.p.).

25   The precise date of the broadcast of "*Die Mietskaserne*" has not been determined though it has been recently suggested that it took place on 12 April 1930 (see Benjamin, 2014 [1930]: 62).

Chapter Three
Resistance and Autonomy

The commune – as both an open form of political cooperation and the direct coexistence of free individuals – may come to represent the only adequate response to our times, namely from individuals who are able to grasp the totality of the problem theoretically and to find new practical answers for a new way of living together in a commune (*eine neue Form des Zusammenlebens*) which would plant the germ cells of the new in the old established equilibrium [...] The fact is that this form of cohabitation is a form of coordinated scientific and human cooperation where scientific work is ultimately conceived as a form of revolutionary practice...

(Dutschke quoted in Chaussy, 1999: 160)

So announced the German activist and member of the West Berlin Socialist German Students Union (*Sozialistischer Deutscher Studentenbund* or SDS), Rudi Dutschke, on 31 December 1966 in a radio interview with Heinrich von Nußbaum on SFB Radio Berlin. Dutschke was responding, in part, to criticism of the SDS in the West German press. Whilst the recently-adopted position of the SDS on the Chinese Cultural Revolution was lampooned for its uncritical adherence to the official Chinese position, Dutschke and fellow West Berlin activists actively sought to remake the notion of the Cultural Revolution "as their own" (Slobodian, 2012: 170; see Klimke, 2010; Von Dirke, 1997). As the historian Quinn Slobodian has argued, students and other activists in West Germany turned to Chinese communism as a key point of reference in the development of a new repertoire of contentious performances and direct action

Metropolitan Preoccupations: The Spatial Politics of Squatting in Berlin, First Edition.
Alexander Vasudevan.
© 2015 John Wiley & Sons, Ltd. Published 2015 by John Wiley & Sons, Ltd.

techniques. For Slobodian, "cultural revolution in West Germany meant a fundamental revaluation of forms of political action and interaction" (2012: 171). And yet, if scholars of radical history have, in this context, begun to explore the breadth and complexity of West German protest cultures from the 1960s onwards (Klimke, 2010; Klimke and Scharloth, 2007; Slobodian, 2012; Thomas, 2003; Varon, 2004; von Hodenberg and Siegfried, 2006), the consensus view of the era known as '68 remains one of national 'liberalisation' and 'democratisation' (Habermas, 1988; see Brand et al., 1986; Kraushaar, 2000; Siegfried, 2006). According to this view, '68 represents the moment when "West Germany began to earn its place among the Western democracies". In other words, the protest movement marked a turning point as traditional norms and values were finally challenged and overturned by a new post-war generation that "sloughed off the residues of fascism and joined the democratic West" (Slobodian, 2012: 5, 6; see also Brown, 2013).

The dominant 'official' story of West Germany's 'cultural revolution' is, of course, much more than one of confiscation and pacification where dissent and radical practice came to be recuperated in the service of capital. Rather, it also signals, if anything, the degree to which today's capitalist Germany is in fact the direct product of the '68 movement's "deepest desires" (Ross, 2002: 6). As Kristin Ross has argued in a related context, "by asserting a teleology of the present, the official story erases those memories of past alternatives that sought or envisioned other outcomes than the one that came to pass" (2002: 6). As a result, we are left with a 'good' version of '68 in which we bear witness to the emergence of a society committed to the principles of democracy and justice. A 'bad' version of '68 – shaped, on the one hand, by escalating violence and extremism, and, on the other hand, by the invention of alternative forms of political activity – is conveniently obscured and forgotten (Slobodian, 2012: 8–9; Ross, 2002: 2).

Much of my own effort in this book is taken up with re-examining the various histories and geographies of these alternatives and their wider significance for how we conceptualise the city as a space of politics. The commune as imagined by Dutschke and others represented, in this context, one important example of the new forms and tactics of protest that distinguished the movements of '68 from their historical antecedents (Klimke and Scharloth, 2007: 5). From teach-ins and happenings to experiments in communal living and working, the action repertoire adopted by the student movement and the extra-parliamentary opposition (*Außerparlamentarische Opposition* or APO) in the late 1960s extended beyond appellative and symbolic expressions of contention and dissent. These new protest techniques were *anticipatory*. They actively prefigured the alternative society that they imagined. It is in this context that I would like to return, in the following, to the figure of the commune and the wider action repertoire of the '68 years. My main aim here is to examine how these practices *endured* and produced new sites of opposition, resistance and autonomy. Whilst it remains something of a commonplace to equate 1968 with the high point of extra-parliamentary opposition in West Germany, scholars have recently shown that the 1970s actually witnessed a significant expansion in the scale and reach of leftist activism (Koenen, 2001; Reichardt and Siegfried, 2010). As this chapter argues, the early history of squatting in West Germany and West Berlin in the late 1960s and early 1970s must be seen as one important product of a wider landscape of protest. It is perhaps surprising, then, that there remains little empirical work on the role of squatting in the emergence and development of the New Left in West Germany. Why, in other words, did thousands of activists choose to break the law and occupy empty flats and other buildings in numerous cities across West Germany? How were practices of shared living and solidarity enacted and articulated across these sites? And in what way did these spatial practices produce an alternative vision of the city as a site of "political action and revolt" (Harvey, 2012: 118–119)?

In order to answer these framing questions, I zoom in on the early development of the squatter scene (*Besetzerszene*) in West Berlin in the 1970s and trace its origins to the changing set of tactics and practices adopted by the student movement and the extra-parliamentary opposition (APO) in the late 1960s (Klimke, 2010). As Belinda Davis (2008) has recently noted, the broad spectrum of New Left activism in West Germany promoted a popular spatial imaginary of protest that situated activism squarely within West Berlin. For many young people, in particular, West Berlin acted as a kind of geographical correlate to a whole host of alternative political activities that shaped and were, in turn, shaped by the city's physical and symbolic fabric (see Scherer, 1984; Scheer and Espert, 1982; MacDougall, 2011b). "Activists," writes Davis, "'made' West Berlin; West Berlin in turn made the activists" (2008: 247). The chapter shows how the performative techniques, skills and dispositions mobilised by various elements of the student movement and the APO in the 1960s provided an important action repertoire that would come to influence the spatial practices of squatters across West Germany and in West Berlin in particular. While these 'direct action' tactics challenged the traditional boundaries between theatrical performance and public space, they were also forced to respond to and assume new forms in the wake of emergency laws that banned public political demonstrations in West Germany in the late 1960s (Kraus, 2007). The chapter goes on to examine early experiments in alternative forms of communal living in West Berlin (Kommune I, Die Bülow-Kommune, Die Anarsch-Kommune) and squatting (Georg von Rauch-Haus, Tommy Weisbecker-Haus) and how the early 1970s bore witness to the production of new 'oppositional geographies' (see Hannah, 2010). The chapter thus situates the emergence of alternative forms of collective living and other self-organised projects as part of a broader turn to the emotional and material geographies of everyday life. Intimate settings – cafés, pubs, alternative presses, bookstores, youth centres, and squatted spaces – offered a radical infrastructure through which alternative support networks were created, friendships made and solidarities secured. At stake here, as the chapter concludes, was the cultivation of alternative political spaces and collective practices that ultimately promoted new ways of thinking about and inhabiting the city.

'The Berlin Commune'

By the time Rudi Dutschke was interviewed on Berlin radio at the end of December 1966, he was no longer actively involved in plans for the development of an alternative commune in West Berlin.[1] In a biography of her late husband, the American writer, activist and theologian, Gretschen Dutschke-Klotz, recalls a conversation in early 1966 in which she and Dutschke discussed the possibility of founding a commune (1996: 96; see also Enzensberger, 2004; Reimann, 2009). A series of meetings followed as activists from across Berlin, including members of the SDS, were invited to attend sessions in which earlier experiments in communal living were explored. This included the work of utopian socialists in 19th-century France and communist activists and educators in post-revolutionary Russia. More recent initiatives in the United States were also discussed. The meetings attracted considerable attention and the number of attendees quickly grew. This came to include, as Dutschke-Klotz recollects, the architect Andreas Reidemeiser who enthusiastically embraced the design of a commune. "We can start to build," he proclaimed, "a new way of living in a house that is purposefully designed for an emancipated future (*für eine emanzipierte Zukunft konzipiert ist*)" (quoted in Dutschke-Klotz, 1996: 97). Reidemeiser drew up preliminary plans for the commune while debate in the meetings shifted to questions of organisation and funding.

In the end, interest in the project was not limited to a dedicated circle of activists in West Berlin. By the mid-1960s, student activists across West Germany had already established a complex network of exchange, organisation and solidarity through which different ideas, policies and tactics circulated (Klimke, 2010). News of the proposed commune soon reached the Munich chapter of the SDS and a small group of activists which included Dieter Kunzelmann. Unlike other chapters of the SDS in West Germany, the Munich group consisted primarily of students with an arts background who, since 1964, had gradually taken over and pushed back the influence of the more traditionalist trade union wing (Klimke, 2010: 55). A group around Kunzelmann, in particular, was also part of Subversive Aktion, a coalition of students and avant-garde artists that were based primarily in Berlin and Munich (Böcklemann and Nagel, 1976). Subversive Aktion was founded in Munich in 1962 by former members of the artist group SPUR which had in effect served as the German section of the Situationist International from 1959 to 1961 (see Kunzelmann, 1998; Lee, 2007, 2011). Subversive Aktion sought to extend SPUR's Situationist repertoire of action-oriented techniques and combine radical political theory with a provocative critique of consumer society that derived its energy from "an international and multi-generational network steeped in anti-fascist resistance, Western Marxism and utopianism" (Lee, 2011: 30).

In this way, the group developed a range of strategies that drew inspiration from the Situationists as well as early avant-garde movements such as French

Surrealism and German Dadaism. As in the case of their Situationist predecessors, at stake here was a mode of critique whose disruptive and experimental character actively worked to imagine an alternative *habitus*. In practical terms, members wrote inflammatory manifestoes and pamphlets, disrupted art openings, organised 'happenings' and were ultimately forced to contest obscenity charges in a Munich court (Lee, 2011; Slobodian, 2012). Recent scholarship has also shown how Kunzelmann and other members of Subversive Aktion, including Rudolf Gasché, turned to the work of the Frankfurt School and Herbert Marcuse in particular as they attempted to develop a more nuanced analysis of modern society. They opened a Berlin section in 1963 and in the following year they recruited Rudi Dutschke and Bernd Rabehl who were students at the Free University. Both Dutschke and Rabehl were themselves close readers of the Frankfurt School and they shared an interest in action-oriented activism and new forms of anti-imperialist struggle (Klimke, 2010; Slobodian, 2012). They were soon involved in attempts by Subversive Aktion to infiltrate and influence the SDS which, by the early 1960s, constituted the most important leftist student organisation in West Germany.[2]

The West Berlin section of Subversive Aktion was renamed the 'Attack Group' (*'Anschlag Gruppe'*) and it soon rose to prominence in protests that accompanied the visit of Congolese Prime Minister Moise Tshombe to West Berlin in 1964. As one historian has pointed out, "West German and foreign students broke the rules of protest together, storming police barricades, bombarding official vehicles with tomatoes and forcing their way into West Berlin's City Hall" (Slobodian, 2012: 51). Dutschke later described the protest as the "beginning of our cultural revolution, in which all previous values and norms were thrown into question, those involved in the action concentrated primarily on themselves, and in action, their self-awareness developed in response to the meaning and goal of the action (*ihre Selbstaufklärung über den Sinn und das Ziel der Aktion weiterführen*)" (Dutschke, 1968: 82). The relative success of the Tshombe protest prompted Dutschke and Rabehl to join the SDS on 27 January 1965 and over the course of the next year they attempted to shift focus within the West German student movement from a moderate line to a new emphasis on direct action techniques and transnational activism.

If the formation of new solidarities between German students and their Third World counterparts has been explored to great effect in a brilliant conspectus by Quinn Slobodian (2012), my main aim here is to pay particular attention to their *spatialisation*. I am supported, in this respect, by the recent work of geographers who have attempted to examine the "spatial relations of solidarity", how they are performed and produced and the different ways in which power, identity and privilege are entangled in these relations (Brown and Yaffe, 2013: 38; see also Davies, 2012; Featherstone, 2012, 2013). For David Featherstone, "solidarity" must be seen as a "transformative relation" in which the very practice of extending solidarity across distance and difference plays an important role in "the active creation of new ways of relating" and "new ways of configuring political relations and spaces" (2012: 5, 6). Featherstone thus shows how acts of solidarity can "reshape the terrain of what is politically possible and what counts or is recognised as political". "This contestation," he argues, "produces new ways of generating political community and different ways of shaping relations between places" (Featherstone, 2012: 7). Solidarities, in other words, are *spatially generative* and by tracing the geographies they create and contest, shape and subvert, a better understanding of their productiveness and endurance may be gained.

It is perhaps not surprising, therefore, that for student activists and other members of the extra-parliamentary opposition in West Germany during the 1960s, political engagement was contingent on the production of new geographies of action and solidarity. What Rudi Dutschke would later describe as a "disposition toward new forms of work, struggle and being with others" was in fact symptomatic of a much wider commitment to direct action, the emotional aspects of protest and new prefigurative modes of social and political organisation (Dutschke, 1981: 8; see Klimke, 2010). It is, moreover, within such a discursive milieu that earlier plans for a commune in Berlin must be located. And it is also, as this chapter argues, out of this very milieu that many elements of the action repertoire later adopted by squatters in West Germany were first assembled and explored. In the end, an understanding of how political action was framed by activists in 1960s' West Germany cannot be separated from how, where and for whom it was actually performed.

For Rudi Dutschke, the plans drawn up for a commune during the early months of 1966 were of a piece with the model of activism that placed supreme importance on aligning *objective* critique with *subjective* forms of revolt (see Dutschke, 1968). In the case of Dutschke, this was a product of a close engagement with Marxism and Critical Theory, most notably the work of Herbert Marcuse who became a talismanic point of reference for many members of the New Left in West Germany during the 1960s. Texts by Marcuse such as *Eros and Civilization* (1955), *One-Dimensional Man* (1964) and *Repressive Tolerance* (1965) helped to foster new alliances between Dutschke and the German SDS and supporters of an emerging American protest movement (Klimke, 2010). At the same time, the formulation of a new model of political action was also based, in no small part, on collaboration between Dutschke, Rabehl and other members of the SDS and Third World students living and studying in West Germany. It is in this particular context that the writings of Frantz Fanon, Che Guevara and other anti-imperialist revolutionaries first took on meaning as key political placeholders for the German student movement (see Slobodian, 2012: 51).[3]

For Kunzelmann and the circle around him, the opportunity to create a commune represented, in contrast, a simple extension of earlier Situationist precepts and the development of a living critique that called for the setting up of "bases for experimental living where people can come together to create their own lives on terrains equipped to their ends" (Kontányi and Vaneigem, 1961). Such Situationist bases would, according to Kunzelmann, act as bridgeheads for the revolutionary transformation of everyday life. If Dutschke advocated a radical realignment of the political *and* the everyday, Kunzelmann favoured, so it would seem, a politics *of* the everyday. In response to Dutschke's plans, therefore, Kunzelmann invited him and other activists from West Berlin to a meeting in the small Bavarian village of Kochel in June 1966 (Kommune 2, 1969). The meeting was attended by the host, Munich SDS member Lothar Menne, and his partner Inge Presser, Kunzelmann and Dutschke, Marion Steffel-Stergar (Kunzelmann's ex-partner), Dagmar Seehuber (later Przytulla, Kunzelmann's partner at the time), and Bernd Rabehl as well as Hans-Joachim Hameister and Horst Kurnitzky who were members of the Berlin '*Anschlag Gruppe*'. Other attendees included Eike Hemmer and his wife Gertrud (Hemmer worked for the SPD politician Harry Ristock) and the Berlin student Fritz Teufel (Dutschke-Klotz, 1996; Enzensberger, 2004; Kätzel, 2002; Reimann, 2009).[4]

Whilst recent recollections of the Kochel meeting have undoubtedly been coloured by personal political commitments and disagreements, retrospective revisions and evasions, it is clear that the meeting represented something of a microcosm for wider debates within the German student movement surrounding the need for collective action and the desire to experiment with a new "repertoire of contention" (Tilly, 1995). As social movement scholars have long argued, movements characteristically mount contentious challenges through disruptive direct action and the collective affirmation of new values (Melucci, 1996; Tarrow, 1998: 5; Tilly, 2008). Action depends, in this case, on a repertoire of practices and strategies that encompass the different "ways that people act together in pursuit of shared interests" (Tilly, 1995: 41). For the activists who met in Kochel, it was a commitment to the work of Herbert Marcuse and its practical implications that ultimately played a key role in bringing them together.

According to Bernd Rabehl (2002), the discussion in Kochel thus focused on the potential for concrete political action and resistance in a capitalist society which was characterised, in Marcuse's view, by its psychological conditioning and its ability to integrate and subsume protest from the outset within a total repressive system (Reimann, 2009; see especially Marcuse, 1964, 1965). For the attendees, the only way out seemed to be a collective break with the traditional forms of everyday life and the concomitant cultivation of new free spaces (*Freiräume*) for leisure, love, reflection and resistance (Rabehl, 2002: 430). Eike Hemmer (1969), writing shortly after the meeting, described the "almost mystical atmosphere" as discussions ran late into the night disconnected from the outside world and focused, so it seemed, on whether the revolutionary moment of '*kairos*' had finally arrived (in Kommune 2, 1969: 17).[5] In the case of Dutschke, this was a question that betrayed perhaps his close reading of key reformist theologians including Karl Barth and Paul Tillich (see Dutschke-Klotz, 1996: 54, 55). It was, however, a question that Dutschke and others – including Rabehl – also connected to an interest in anti-imperialist liberation movements and the writings of Fanon and Guevara. Kunzelmann, on the other hand, was convinced that new forms of protest offered the possibility for collective self-transformation and an opportunity to expose the repressive psychological dispositions of other group members which, Kunzelmann believed, prevented effective action (Reimann, 2009: 129; see also Kommune 2, 1969). In the end, as Kunzelmann's ex-partner Dagmar Przytulla (formerly Seehuber) recollected, the meeting "was mainly a discussion between Kunzelmann and Dutschke, and it went back to the question of whether political actions could be formed and sustained without living together on an everyday basis." "Communes should therefore be created," she noted, "and, if I remember right, the idea was Dutschke's" (Przytulla in Kätzel, 2002: 205).

Dutschke quickly returned to West Berlin after the meeting and he was soon followed by Kunzelmann and other members of the Munich SDS in September 1966 who were keen to pursue plans for the setting up of a commune. Dutschke-Klotz was less than complementary of Kunzelmann and his 'ambitions'. "He will destroy everything," she told Dutschke. In her view, "the [original] project as the expression of a new form-of-life under revolutionary living conditions did not fit in with Kunzelmann's imagination." "He limited himself," so Dutschke-Klotz opined, "to the demand for sexual freedom, emotional detachment and the use of 'psychological terror'" (1996: 97). Yet Kunzelmann's psychoanalytical approach to collective action and group therapy retained considerable support within the wider working group which had grown to over 50 members in the wake of the annual SDS conference in Frankfurt (Kommune 2, 1969). On 26 November 1966, a group of activists including Kunzelmann, Hemmer and the SDS member Rainer Langhans stormed into a student meeting with the president of the Free University, Hans-Joachim Lieber. Hemmer seized a microphone from university administrators and read out the text of a flyer that condemned the University. "We have to cope with poor working conditions," it said, "miserable lectures, stupefying seminars, and absurd exam requirements. When we decline to be trained by narrow-minded experts (*Fachidioten*) to become ourselves narrow-minded experts, we run the risk of ending our study without degrees". The group urged students to quit university, find a job in a factory, buy a house and turn it into a commune for 'free love' and a base for radical action (see also Slobodian, 2012: 177).[6] As Bernd Rabehl proclaimed with great excitement in the same month, "our goal is to set up a commune. The establishment of a commune is the necessary condition for our practice (*Unser Ziel ist das Setzen der Kommune. Setzen der Kommune ist die Voraussetzung von Praxis*)." What Rabehl described as a form of "anarchist practice" represented, he argued, a "disruption of theory." "We have decided," he added, "to move on from an analysis of tendencies and trends. It is *practice* that matters in *this* moment. The last anarchist movements failed because it was not the right time for action. Historically, and for the first time, there is a way for us" (Rabehl quoted in Kommune 2, 1969: 19; emphasis added).

Neither Dutschke nor his wife shared Rabehl's excitement. By November, they had already distanced themselves from the working group that they had originally set up and, according to Kunzelmann, were now focused on the founding of a radical "scientific commune (*Wissenschaftskommune*)" (Kunzelmann, 1998: 59). On the very day that Dutschke spoke on SFB radio, a meeting was held by Kunzelmann and other activists, and it was decided in the early hours of the morning of 1 January 1967, that the establishment of a commune – what later became known as Kommune I – would go ahead (Enzensberger, 2004: 96–97; see Kunzelmann, 1998). At this point, a number of activists backed out. Many others still lived in tiny bedsits and while the SDS announced their own plans to support the commune in February 1967, little progress was made. It is against a background of increasing inertia that a small core of activists decided to move into a workshop in the leafy Berlin suburb of Friedenau on 17 February 1967. The workshop, which belonged to the writer Uwe Johnson, had been sublet to Ulrich Enzensberger, brother of the critic and poet Hans-Magnus Enzensberger. As Kunzelmann, Seehuber, Teufel, and Detlef Michel moved into Johnson's workshop at Niedstraße 14, another group – including Ulrich Enzensberger, Hans-Joachim Hameister and Dorothea Ridder – moved into Hans-Magnus Enzensberger's empty flat a couple of blocks away on Fregestraße. It was only in May of the same year that members of Kommune I were able to move into their own shared apartment at Kaiser-Friedrich-Straße 54a in the Berlin district of Charlottenburg.

The founding of Kommune I has acquired something of a mythical status within a broader history of the extra-parliamentary opposition in West Germany in the late 1960s. If press coverage at the time was largely caricatural in tone, mingling scandalous biographical detail with casual impressionistic accounts of commune life, recent scholarship has done little to challenge this one-dimensional portrayal. We are still reminded of the commune's "alternative hippie lifestyle" which included "long hair, unkempt and colourful clothing, drug-taking, free love, collective nudity, and so on" (Thomas, 2003: 99). At the same time, the commune's *political* activities have often been reduced to a collection of colourful episodes and anecdotes that were of a piece with the commune's putative bohemian status. To be sure, this is a self-image that was, in many respects, cultivated by members of Kommune I itself as their political tactics and ensuing court cases guaranteed celebrity treatment. And yet, the very repertoire of contention produced by the commune *did* matter and, as I will show in the remainder of this chapter, played a central role in the development of new protest cultures in West Germany from the late 1960s onwards. At stake here are a series of practices that would help to create the necessary conditions for the emergence of urban squatting as a radical social movement in cities across West Germany in the 1970s. This is not to say that the life and structure of a Berlin commune in the 1960s was identical to that of an illegally squatted apartment in Kreuzberg or Schöneberg in the late 1970s. Nor is it to suggest that the history of protest in West Germany and West Berlin can be plotted as a strongly vectored narrative that culminates in the actions of squatters in the late 1970s and early 1980s. Rather, it is, if anything, to insist on an elective affinity between the practices of squatters and the new forms of action and solidarity that were *immanent* to the tactics developed by the countercultural Left in Germany as it emerged in the 1960s. In order to do so, I argue that we need to focus in particular on the city as a key laboratory of protest and revolt. This was a widely-held view at the time, one shared by the activist, poet and former political commentator for the BBC German service, Erich Fried, who addressed a large international audience of activists and students at the Vietnam Congress in Berlin in February 1968 arguing that "big cities are a potential forum (*Die grossen Städte sind ein potentielles Forum*)." "The city is not only our own forum," Fried conceded, "it is also the forum of our opponents and even more so, as long they rule over us. For whom the forum belongs is a question of power which is, in turn, a question of resistance and struggle...In this way, cities are key sites for us (*Dann sind die Städte unsere Schlüsselstellungen*)" (quoted in Dutschke-Klotz, 1996: 185).

Performing the Political

The cardinal significance of the city as both the setting and stage for new forms of contentious collective action has remained something of a blind spot within recent scholarship on radical politics and the emergence of the New Left in West Germany in the 1960s (exceptions include MacDougall, 2011b; Vasudevan, 2011a, 2014a). In the case of West Berlin, the struggle for self-defined autonomous spaces has a long and contested history that can be traced back, as this chapter shows, to the late 1960s and the development of communes and other collective spaces for political action. At the same time, the emergence of these alternative urbanisms is also, as I argued in Chapter 2, a product of various forms of "resistance to urban renewal as an idea and practice" (MacDougall, 2011b: 155). By the 1960s, a new generation of professional planners and architects had joined with community organisers, tenant activists and other local citizens' initiatives with a view to highlighting and reversing the worst excesses of post-war urban redevelopment and a desire to assemble alternative forms of urban living and working. That this would later drive autonomous anti-capitalist struggles in Berlin which focused on housing shortages, property speculation and social inequality is, of course, a main theme of this book. Activism of this kind would reveal what urban geographers have long argued, namely that space is a product of continuous conflict and struggle and that the city is "fundamentally shaped and altered by these multi-layered political and cultural contests concerning the use and definition of urban space" (MacDougall, 2011b: 156; see Harvey, 2008, 2012; Brenner et al., 2011; Smith, 1996).

If attempts to redefine and occupy the urban thus played a central role in the practices of squatters, the tactics they adopted represented a productive re-imagining of the action repertoire that accompanied new forms of political protest in West Germany in the 1960s. As the repertory motif furthermore suggests, such forms of political contestation were themselves *performances*. 'Performance' was, in this context, far more than a conceptual apparatus or a theoretical shibboleth for describing a politics of collective action. This is a view shared, in turn, by a number of theatre and performance historians who have recently argued that the emergence of an alternative public sphere in West Germany in the late 1960s relied, in no small part, on a new range of tactics whose provenance can be attributed to the performing arts (Boyle, 2012; Gilcher-Holtey, Kraus, and Schöler, 2006; Kraus, 2007; Papenbrock, 2007). Agitprop, happenings, teach-ins and street theatre contributed to the formation of a 'new political pedagogy' and testified, as Dorothea Kraus (2007) has shown, to a performative milieu characterised by the blurring of boundaries between the traditional place of theatrical performance and public space. To perform the political was, in other words, to produce geographies of collective dissensus that were composed through and by these practices.

It against this backdrop that members of Kommune I turned to a range of performative techniques that ultimately came to occupy a central role in the early history of the commune. The techniques adopted what Anja Kanngieser (2012, 2013) has more recently described as the "performative encounter" as a strategy for facilitating new ways of making alternative political worlds. According to Kanngieser, such an encounter emphasises the desire "to move through and beyond political circles, to work on issues that affect people on a day-to-day basis and to participate in self-determined and shared struggles". As Kanngieser argues, this is an encounter that draws on "creative political praxes that take as their prerogative the disruption of the borders between artist and audience, activist and non-activist, politics and everyday life, amateur and professional, alternative and mainstream" (2012: 286). In the case of Kommune I, such an encounter built on a repertoire of tactics that treated 'performance' as a political *and* aesthetic event. These were tactics that emerged out of an artistic avant-garde milieu that encompassed most notably the Situationists and their German counterparts and placed particular emphasis on the radical transformation of everyday life. In the words of Fritz Teufel, one of its founding members, Kommune I represented a "revolution of everyday life, an abolition of private property, a breaking of the achievement principle, a proclamation of the pleasure principle (*Revolutionierung des Alltags, Abschaffung des Privateigentums, Brechung des Leistungsprinzips, Proklamation des Lustprinzips*)" (Teufel in von Uslar, 1997: 72). Rainer Langhans defined the commune, in turn, as a "political group with a shared living area" while Kunzelmann spoke of a "political group that seeks to combine political and human existence".[7] "We took the line," added Ulrich Enzensberger, "that a fulfilled individual life of freedom and the freeing of desires is directly linked to meaningful political work" (Enzensberger quoted in Chaussy, 1999: 137).

For members of Kommune I, what constituted 'political work' was never confined to the interior space of the commune. Rather, the development of a radical mode of political action depended, in the first instance, on a set of performances that not only took place *in* public but were in fact constitutive of new *public geographies* of dissent and resistance. As Subversive Aktion, one of the commune's direct antecedents, proclaimed as early as 1963, "critique must turn into action. It is action that exposes the reign of oppression (*Kritik muss in Aktion umschlagen. Aktion entlarvt die Herrschaft der Unterdrückung)*" (quoted in Kraus, 2007: 112). In the case of Subversive Aktion, 'action' began as a rehearsal and recasting of existing artistic and political practices that were first developed in the late 1950s by the Situationist International. The Situationist International (SI) was formed in 1957 as the fusion of a number of small avant-garde groups in Europe that included the Lettrist International, the International Movement for an Imaginist Bauhaus and the London Psychogeographical Association. Prominent members included Guy Debord, Asger Jorn and Michelle Bernstein. If the overarching aim of the group was the development of a total critique of modern society, this was to be achieved, in part, through the development and performance of 'situations' designed to challenge and alter "the meaning of social interactions and to allow for new perspectives and experiences" (Klimke, 2010: 55). These were *spatial* practices that represented a process of critical questioning forged out of a radical commitment to transforming social space and everyday life (see Pinder, 2005: 129). This was, moreover, a commitment to rethinking the city as a site of artistic and political action. As the geographer David Pinder reminds us, the Situationists "recognised cities to be key sites in the reproduction of social relations of domination as spaces of alienation and control". At the same time, they also acknowledged, according to Pinder, "the possibilities that lay embedded within these environments as they viewed cities as potential realms of freedom through which people could transcend alienation and create spaces in keeping with their own needs and desires, thereby realising their true selves as living subjects" (Pinder, 2005: 128).

Scholars of radical history in West Germany have thus shown that groups such as Subversive Aktion, Kommune I, and the wider extra-parliamentary opposition of which they were a part, combined an action repertoire that borrowed heavily from the Situationists with a set of tactics that were inspired by the American New Left and liberation movements in the Third World (Höhn, 2002; Klimke, 2010; Kraus, 2007; Lönnendonker et al., 2002; Slobodian, 2012; Varon, 2004). If Dutschke, Kunzelmann and others sought to translate an analysis of critical theory into political action, this was increasingly predicated, they argued, on the production of new subjectivities and an emphasis on the "subjective aspects of revolt" (Slobodian, 2012: 59). In Dutschke's own words, "it is only because these actions change us *on the inside*, that they are political (*Weil uns diese Aktionen innerlich verändern, sind sie politisch)*" (1968: 76; emphasis added). "The permanent revolution," Dutschke insisted, "is impossible without the development of a new subject" (1968: 77). The struggle for radical social change seemed, in other words, unthinkable *without* a process of subjective transformation. "Breaking through managed consciousness (*verwaltes Bewusstsein)*," Dutschke added, "was the precondition of liberation"

(Dutschke reprinted in Böcklemann and Nagel, 1976: 327).[8] As the former activist Peter Schneider concluded, writing in the journal *Kursbuch* in 1969, the first step in the struggle was to use "methods of direct action and agitation" to produce a "revolutionary consciousness" (1969: 1–2).

While these assertions seem bold and startling, these were, of course, sentiments that would come to reflect a protest movement that only grew in its militancy during the late 1960s and 1970s. At the same time, they also encompassed a form of political action that increasingly sought to translate an understanding of "revolutionary theory" into a tangible "sensory experience" (Dutschke quoted in Slobodian, 2012: 73). At the SDS conference in 1967, Rudi Dutschke and Hans-Jürgen Krahl went so far as to argue that there was a need to confront "executive state force" in order to expose the "abstract violence of the system" as a "sensory certainty (*sinnlichen Gewißheit*)".[9] Quoting Che Guevara, Dutschke and Krahl argued that "the 'propaganda of the bullet' in the Third World must be completed by the 'propaganda of the deed' in the metropoles" (quoted in Dutschke-Klotz, 1996: 150). For many West German radical students, the speech highlighted the importance of the Third World 'guerrilla' as an archetype for the militant activist. It also signalled a further consolidation of the 'anti-authoritarian faction' within the SDS in West Berlin and across West Germany and the emergence of new *oppositional geographies* as activists enthusiastically embraced direct action and resistance.

In so doing, a new repertoire of contention emerged that built on tactics developed by the West German SDS in the mid 1960s. According to the historian Martin Klimke, these were civil disobedience tactics still marked by a "limited but open and symbolic overstepping of rules, laws and values" that took shape in the form of direct actions such as sit-ins or teach-ins (2010: 53).[10] The uptake of these tactics coincided, however, with the emergence of Subversive Aktion and its growing role within the SDS. Over the course of 1965, they drew support away from the moderate wing of the West Berlin SDS and toward a new emphasis on political confrontation and artistic provocation. By the spring of 1966, the result, as Ingrid Gilcher-Holtey has argued, was an inversion of the relationship between theory and practice with "new forms of action now preceding theoretical debate" (1998: 180). A decisive factor was the start of the US bombing campaign in North Vietnam which played a key role in radicalising students who expressed growing scepticism towards the viability of an ostensibly "liberal public sphere" as a medium for articulating anger and expressing dissent (Slobodian, 2012: 74). It is, moreover, against this very backdrop that the activities of Kommune I ultimately came to occupy a central role within the anti-authoritarian new Left in West Berlin.

On 10 December 1966, just a few weeks before the official founding of Kommune I, a protest against the Vietnam War took place in West Berlin on the Kurfürstendamm which was the main shopping avenue in the Western half of the city. A number of protesters left the officially approved route and were chased by the police and arrested. After the final rally, the situation escalated as future members of Kommune I initiated what would later become known as the 'Christmas Happening' ('*Weihnachtshappening*') in front of a café on the Kurfüstendamm. The protesters performed Christmas songs and a Christmas tree decorated with American flags was set on fire. The police intervened and 63 arrests were made (see Kraus, 2007). The following week, the Berlin chapter of the SDS organised a protest in response to what they described as "the brutal enforcers of this democracy (*die brutale Schläger dieser Demokratie*)" (pamphlet quoted in Dressen, Kunzelmann and Siepmann, 1991: 201). The 'walking demo' ('*Spaziergangsdemonstration*') was not a traditional protest march and represented, if anything, a new alignment of urban geography and direct action. Over 200 activists met in front of Café Zuntz on the Kurfürstendamm on 17 December 1966 at 3 o'clock and dispersed into small groups of two to three distributing pamphlets to passers-by explaining the reasons and tactics of the protest walk. "As walkers, we meet at predetermined points," explained the pamphlet, "surrounded by the friends of the police. We aim to disrupt, then to disappear and become pedestrians and re-emerge at a different location" (quoted in Dressen, Kunzelmann and Siepmann, 1991: 201). Protesters and shoppers were thus indistinguishable, prompting a brutal crackdown by the police. The police rushed into the crowded sidewalk wielding batons indiscriminately. Eighty-seven arrests were made including numerous bystanders and one off-duty police officer (Kraus, 2007: 171). For the organisers, the protest had the desired effect. As a spontaneous performance, it transformed, they argued, the 'system's' abstract mechanisms of oppression into material certainties.

The development of new public geographies of protest by student activists and members of Kommune I represented, at the same time, a highly spatialised response to growing anxieties about proposals for new Emergency Laws (*Notstandsgesetze*) which would allow the government to suspend civil liberties in the event of a natural disaster, invasion or general strike (Thomas, 2003). The idea of amending the West German constitution was first suggested in 1958 though the *Bundesrat* raised objections to the use of the new laws to resolve industrial disputes and in 1960 the *Bundestag* rejected the bill and it was sent back to the committee stage (Thomas, 2003: 88). In 1965, the CDU tabled a new bill whose enactment would allow for censorship of the press, "a ban on freedom assembly or freedom of forming unions or associations, a removal of the right to choose one's employment, withdrawal of legal guarantees of personal freedoms, as well as the ability to introduce forced labour and to deploy the police to resolve internal disputes" (Thomas, 2003: 88). For many, the proposed Emergency Laws raised once again the spectre of authoritarianism and the state of exception under which the National Socialists governed between 1933 and 1945. Unions launched a campaign against the Emergency Laws and were later joined by student activists and other citizen initiatives across West Germany.

The *publicness* of student protests in West Berlin in the late 1960s must be understood, in part, as a recognition of the importance of different public sites (the university campus, the courtroom and the street) for *performing* political alternatives in the face of potential suppression. In the case of Kommune I, protest also represented a desire to blur the boundaries between private social space and public political space, between stage and street, between an individual desire for creative spontaneity and a collective need for committed action (Merrifield, 2013a). Early accounts of Kommune I drew attention, in this way, to both the internal workings of the commune and the public activities of its members. This was clearly visible, for example, during the April 1967 visit of US Vice-President Hubert Humphrey to West Berlin. While students at the Free University planned a demonstration against the Vietnam War, members of Kommune I aimed to target Humphrey's entourage during a visit to a gallery in the Schloss Charlottenburg (Thomas, 2003: 99). The night before Humphrey's visit, West Berlin police raided the Kommune I residence in the belief that an assassination attempt was being planned. Eleven students were arrested including eight members of Kommune I. The police released details to the local tabloid press who were quick to report a successful police operation that had thwarted an attempt on Humphrey's life supported, so they alleged, by 'bombs' and other material from Peking. Only a day later, the students were released after pudding, yogurt, and ingredients for butter tart were found in the residence instead of the suspected explosives (see Klimke, 2010: 156; Köhler, 1967: 61; Lönnendonker and Fichter, 1975: 151).[11] As members of Kommune I explained, their intention was to throw cakes at the Vice President, a provocative action inspired by activists in the Dutch 'provo' movement.

The 'pudding assassination attempt' ('*Pudding Bomben Attentat*'), as it became know, served to reinforce the status of Kommune I as protest provocateurs par excellence. Only a month later, members of Kommune I were expelled from the SDS after they distributed a series of unauthorised flyers at the Free University in Berlin signed 'SDS'. The flyers insulted university administrators, accused student leaders of spineless careerism and threatened arson on campus (Lönnendonker et al., 2002: 455; Slobodian, 2012: 103). A few weeks later, a new series of flyers on the conflict in Vietnam were distributed once again on campus. The sequence of flyers drew attention to a recent fire in a Brussels department store which had left over 300 people dead. The first flyer described how the fire was started by a group of Vietnam activists and proclaimed that groups in other cities might be encouraged to use the power of "this big happening (*dieses Großhappenings*)" in similar actions. The second flyer argued that a "burning store with burning people conveys for the first time in a major European city the crackling sensation of Vietnam – being there and burning, too – that we have so far missed in Berlin". The next flyer was titled "When will Berlin department stores burn?" ("*Wann brennen die Berliner Kaufhäuser?*") and concluded with the following text:

> If fires erupt in the near future, if barracks blow up somewhere, if the grandstands in a stadium collapse, don't be surprised. No more so than when the American marched over the Line of Demarcation, when Hanoi's city centre was bombed, or when the marines invaded Chica. Brussels gave us the only answer to that: burn, ware-house, burn![12]

In the volatile atmosphere that followed the shooting of the student Benno Ohnesorg, the police and judiciary seized their chance and, on 9 June 1967, the district attorney's office charged Kommune I members Fritz Teufel and Rainer Langhans with "inciting life-threatening arson (*Aufforderung zur meschengefährdenden Brandstiftung*)."[13]

The killing of first-time protester Benno Ohnesorg by West Berlin police during demonstrations against the visit of the Shah of Iran on 2 June 1967 marked indeed a major turning point for the student movement and the broader extra-parliamentary opposition (see Fahlenbrach, 2002; Thomas, 2002; Soukup, 2007). For many protestors, it heralded a sense of crisis and a further examination of the relationship between the use of violence and the nature of the West German state. While historical scholarship has focused on the significance of 2 June for the West German student movement, less attention has been paid to the transnational dimensions of the protest. After all, German students and Kommune I, in particular, participated in the first instance as a show of solidarity with Iranian dissidents in West Germany and, as Quinn Slobodian (2012) has pointed out, the West Berlin police responded violently to both German *and* Iranian students who had chosen to confront the Shah with a day of protests. While the first confrontation between protesters, the police and the SAVAK (Iranian secret service) took place in front of the Schöneberg City Hall in the early afternoon, the main conflict came in the evening. Protesters were penned – *kettled* as we would now say – into a closed lane on the Kaiserdamm in front of the *Deutsche Oper* and were savagely beaten by police. As the Shah watched a performance of Mozart's *The Magic Flute*, the police chased and battered demonstrators. When protesters attempted to flee, they "barraged with water cannons". In the ensuing mêlée, 39-year-old plainclothes officer Karl-Heinz Kurras shot 26-year-old literature student Benno Ohnesorg in the back of the head (Slobodian, 2012: 116; see Soukup, 2007: 39–75 for a full account). Ohnesorg later died in Moabit Hospital as pitched battles between police and protesters continued into the night.[14]

It is undoubtedly true that the shooting served as a catalyst for the further radicalisation of the student movement. According to Nick Thomas (2003: 107), "it is not too extravagant to say that Ohnesorg's death changed the political landscape irrevocably". Katrin Fahlenbrach (2002), in turn, described 2 June as a "critical event" whilst others heralded it as the 'real' beginning of West Germany's '1968' (168; see Fichter and Lönnendonker, 1977). And yet, as this chapter argues, it is also important to challenge the prevailing mythologising of 1968 and the post-1968 era which has largely focused on either (1) the retreat of the student movement from active engagement in a public political sphere to various narcissistic experiments with 'life-style' politics; and/or (2) the further radicalisation of particular groups who began to move, as Belinda Davis (2006) has pointed out from, "'guerrilla theatre' to 'urban guerrilla' tactics", from the provocative staging of symbolic violence to violent activism culminating in the actions of the *Rote Armee Faktion* (Red Army Faction) and the *2. Juni Bewegung* (June 2nd

Movement) (224; see McCormick, 1991; Weinhauer et al. 2006). These narratives tend to occlude the complex relationship between contentious politics, direct action and the various geographies of protest and dissent that emerged in West Germany during the 1960s. It is in this context that I turn in the remainder of the chapter to the relationship between experiments in communal living in West Berlin (Kommune I, Die Bülow-Kommune, Die Anarsch-Kommune), early attempts at urban squatting in the same city (Georg von Rauch-Haus, Tommy Weisbecker-Haus) and wider developments within the extra-parliamentary opposition. The creation of "liberated zones" in communal apartments, occupied university lecture halls and spontaneous demonstrations and happenings provided, I argue, a template for alternative forms of collective living and other self-organised projects (Slobodian, 2012: 16). The early 1970s bore witness, in this way, to the production of new oppositional practices and a *radical infrastructure* through which autonomous spaces were created and practices of self-determination and resistance were developed, shared and secured.

From Protest to Occupation

For many students and other left-leaning Germans, the events of 2 June provided conclusive evidence that the West German state was willing to answer protest and dissent with deadly violence. Students were also outraged by distorted coverage in the local press which defended the police and blamed protesters for inciting violence. It is therefore not surprising perhaps that Ohnesorg's death proved decisive in the scaling-up of the student movement and the extra-parliamentary opposition. On 9 June, over 5000 FU students (roughly 50 percent of the student body) participated in a memorial procession for Ohnesorg in West Berlin after which his body was driven through East Germany to his hometown of Hannover by a convoy of more than 100 automobiles (Slobodian, 2012: 117, 118). The following day, around 10,000 students from all over West Germany took part in a memorial march through the streets of Hannover. According to the historian Nick Thomas, up to 200,000 students participated in demonstrations across West Germany in the days immediately following 2 June (2003: 114). A survey in *Der Spiegel* in turn found that over 65 percent of students had been politicised by Ohnesorg's death.[15]

For others already active in student politics and other networks of protest and solidarity, the events of 2 June only served to intensify and further radicalise the use of direct action techniques. The need for escalation had, in fact, already been voiced to great effect by the student and activist Peter Schneider during an earlier sit-in to protest at the Vietnam War on 5 May 1967 at the FU:

> We provided information in all objectivity about the war in Vietnam, although we have found that we could quote the most unbelievable details about the American policy in Vietnam without this triggering our neighbours' imagination, but that we

only needed to step on a lawn where this is prohibited to cause honest, general and lasting terror [...] That's when we came up with the idea that we must first destroy the lawn before we can destroy the lies about Vietnam [...] that we must first break the rules of conduct for the building before we can break the rules of the university (*daß wir erst die Hausordnung brechen müssen, bevor wir die Universitätsordnung brechen können*) [...] That's when we realised that in these prohibitions, the criminal indifference of a whole nation runs rampant (Schneider quoted in Larsson, 1967: 162).

To Schneider and other activists, protest *in* West Germany was already part of a wider anti-imperialist struggle that linked students to revolutionaries in the so-called Third World. Many therefore understood and experienced 2 June as the concrete expression of transnational networks of "activism and state repression that linked Iran and West Germany" (Slobodian, 2012: 133).[16] It is, moreover, in this very context that the action repertoire adopted by some West German New Leftists took on an even greater militancy. In September 1967, for example, Rudi Dutschke and members of Kommune I led several hundred participants of the national convention of the German SDS to America House in Frankfurt. Protesters invaded the House, disrupting a panel discussion on US intervention in Vietnam. As a report produced by the consulate noted, "students 'captured' the platform, raised Viet Cong flags, led in singing of international[e] and chanting slogans 'Ho Chi Minh', 'Amis, CIA – go home'" (quoted in Klimke, 2010: 189). A couple of months later, Dutschke, the Iranian activist Bahman Nirumand and others stormed the Kaiser Wilhelm Church on the Kürfurstendamm in West Berlin during a Christmas service. Churchgoers attacked the protesters and Dutschke had to be taken to hospital to be treated for the injuries he sustained (Slobodian, 2012: 164; Dutschke-Klotz, 1996: 171–172).

Others went even further. In the immediate aftermath of the Ohnesorg shooting, a mass meeting was held at SDS headquarters in West Berlin. One attendee, Gudrun Ensslin, proclaimed that "they are trying to destroy us. We must protect ourselves. We must arm ourselves" (Becker, 1977: 41). While Ensslin would later become a key member of the Red Army Faction (*Rote Armee Faktion*, RAF), she and other activists, including Andreas Baader, Thorwald Proll and Horst Söhnlein, were already resorting to more extreme measures. On the night of 3 April 1968, the group planted incendiary devices in two Frankfurt department stores which caused over 2 million DM in damages. The following Friday, Rudi Dutschke was shot and seriously wounded by a far-right sympathiser in West Berlin. Protests and violence erupted in the city and across West Germany. Wolfgang Kraushaar estimates that over 300,000 participated in protests on the day after the shooting (1998: 108). In West Berlin, students targeted the office of the Springer publishing house which they believed was responsible for inciting violence against the student movement and the wider extra-parliamentary opposition. A group of protesters led by Bernd Rabehl charged and broke through police lines and briefly occupied the building. The car park was also set on fire. According to Nick Thomas, "large, sustained and violent protests took place throughout the weekend in at least 20 cities, with barricades being constructed outside Springer plants and offices in Hamburg, Essen, Esslingen, Frankfurt, Cologne and Munich" (2003: 173). The widespread protests were the largest since the interwar period.

And yet, the spectre of violent protest and resistance that accompanied elements of the extra-parliamentary Left in West Germany and culminated with the rise of the RAF and the events of the German Autumn (*Deutsche Herbst*) in 1977 has tended to overdetermine how the post-68 era has been narrated and received. These developments, as scholars of new social movements remind us, were part of a complex landscape of contentious politics that encompassed a range of tactics and practices (Melucci, 1988; Tarrow, 1989). If anything, the scale and depth of leftist activism in West Germany *grew* in the late 1960s and 1970s and increasingly turned to the development of spaces where people could escape the proscriptions of public protest and come together to imagine alternative worlds and live and work autonomously. At stake here was a commitment to producing enduring sites that served as bases for political action and spaces for daily life (Feigenbaum, Frenzel and McCurdy, 2013: 2).

The court case of Kommune I members, Fritz Teufel and Rainer Langhans, which began on 6 July 1967 before the Sixth Criminal Division of the District Court in Berlin-Moabit should be seen in this wider context. Teufel and Langhans used the trial as a public platform to apply and refine their repertoire of anti-authoritarian and Situationist practices. Before an 'audience' of over 80 students and 60 members of the press, the space of the court became – quite literally – a 'theatre of protest' as Teufel and Langhans challenged the traditional roles conferred to them in a formalised legal environment (Klimke, 2010: 266; Davis, 2008). They repeatedly addressed the public gallery, questioned the prosecutor and judge and expressed reservations about the legitimacy of the legal system and the charges brought against them. What became known as the 'Moabit Soap Opera' ('*Moabit Seifenoper*') turned the courtroom into a stage where Teufel and Langhans could 'perform' as they pleased while ridiculing the justice system and exposing its authoritarian nature (see Scharloth, 2010).[17] In Langhans' (1968) own words as he awaited sentencing, "we don't often get to see such a play. No author of a play for the theatre of the absurd could have come up with a better one. Often we were not even players in it, because it was not our game. It would not have even occurred to us that one could create plays like this. We only became players and gradually more like directors when we realised the opportunities being offered to us [...] We are curious [about the sentence] and thank you for this play" (Langhans and Teufel, 1968: passim). In the end, neither Teufel nor Langhans were convicted while the 'Moabit Soap Opera' received extensive press coverage leading one German theatre critic to praise Teufel as the "most remarkable German playwright of the sixties" (quoted in Carini, 2003: 68).

The turn to 'performance' as a protest tactic adopted by groups such as Kommune I played, as this chapter shows, an important role in the development of new social movements in West Germany in the 1960s and 1970s. At the same time and according to a number of historians, the performing arts were themselves undergoing a period of politicisation and radical transformation (Gilcher-Holtey, Kraus and Scholer, 2006; Kraus, 2007). While these developments point to the production of critical counter-geographies that sought, in the first instance, to reclaim and sustain a healthy public sphere, for members of Kommune I and other activists they were also linked to the cultivation of more intimate protest geographies. The 'Moabit Soap Opera' represented, in many respects, a high point for Kommune I and confirmed the status of Teufel and Langhans as protest 'pop stars' and the group as 'experts' in the re-imaging of public protest in West Germany and West Berlin in particular (Reimann, 2009). The group's commitment to communal living should not, however, be overlooked and it represented an equally significant aspect of their action repertoire. The famous photograph of Kommune I that was distributed in a promotional brochure produced by the group and later published in the magazine *Der Spiegel* in June 1967 only served to reinforce their self-professed desire to challenge the boundaries of familial, sexual and gender relations (see Herzog, 2005). The photograph, as Dagmar Przytulla (née Seehuber), one of the founder members of the commune, recollects, was meant "to break sexual taboos [...] and portray the commune as a group that overstepped the moral boundaries set by society" (quoted in Kätzel, 2002: 214).

The breaking of taboos took on, of course, an immediate spatial form. Recalling his first visit to Kommune I in May 1967, the activist Peter Schneider described the main room as "a manifesto in furniture". "It announced," he continued, "the cancellation of the standard division of labour that we find in the 'bourgeois apartment' (*es verkündete die Aufhebung jedweder Arbeitsteilung in der 'bürgerlichen Wohnung'*). Books, packages, dishes, chairs, desks, beds, everything was very confused. The room only lacked a toilet bowl" (2010: 134). The apartment shared by members of Kommune I at Kaiser-Friedrich-Straße 54a was, in fact, a large 6-and-a-half-room apartment on the third floor of a block that was primarily used as a brothel. According to one of the original members of the commune, the flat was organised around two large rooms. One was used as a library, the other as a workroom and dormitory with mattresses spread on the floor. There were two smaller rooms and a room originally used by servants. Most of the 'communards' slept in the large room that served as the main dormitory while others worked around them on political posters and pamphlets (Kätzel, 2002: 208).

In the summer of 1968, the commune moved into an empty warehouse at Stephanstraße 60 in Berlin-Moabit and began renovations of the 3-storey building. Eventually, the three floors were divided between a discotheque, a floor for visitors and the main space of the commune. The move, if nothing else, served to highlight the growing divisions between various members of Kommune I.

As public happenings and interventions became less frequent, some members, most notably Rainer Langhans, advocated a shift away from agitprop provocation, histrionic protest and the strident militancy increasingly adopted by certain factions in the extra-parliamentary opposition. What the historian Aribert Reimann (2009: 195) described as the "pop-cultural turn" of Kommune I heralded a further re-orientation of the commune around a lifestyle politics that drew inspiration from a wider network of subcultural practices (DIY publishing, underground music, drug use, etc.). As one publication produced by the Kommune proclaimed as early as February 1968, "we want to open a restaurant, with music and everything, even movies and meetings and of course lots of light and everyone should dance. It should be a center, a meeting place, where you feel at home, you get to know yourself and other people you want to do something with" (quoted in Reimann, 2009: 201).

And yet, as much as Kommune I cultivated a myth of shared communal living and 'free sexuality', for many women, the reality that remained was one of patriarchal dominance and repression (Amantine, 2011: 71; Kätzel, 2002; see also Berndt, 1969). According to Dagmar Przytulla, Kommune I "was no place for women who wanted to live freely and in solidarity with men outside conventional social norms." "I got involved in the commune," she continued, "because I thought it might be a different kind of society, where women could exercise their rights naturally, but I was disappointed. The patriarchal structure of society was actually reinforced in the commune even though something completely different was written on its banner. Men blustered about how we all had to change something in ourselves. Although they had the concept in theory, in action, they were far from it." Przytulla's observations were supported by a former communard who "noticed how men often felt 'oppressed' by housework and that they all had important appointments and meetings. Most of the time it was the women who had to deal with it." "The crazy thing," as another activist recalled, "is that it was indeed the men who instigated the revolution though they themselves underwent no process of emancipation unlike most of the women" (Kätzel, 2002: 130, 253).

While disagreements and tensions ultimately contributed to the dissolution of Kommune I in November 1969, they also drew attention, more generally, to the role of micropolitical practices in the life cycle of activist communities (see Brown and Yaffe, 2013; Davies, 2012; Featherstone, 2012; Gould, 2009). As the sociologist Lynn Owens (2008) reminds us, "social movements are more than the sum of their actions. They are a cluster of unstable and evolving identities [...] To act is not just to do something, it is to *be* something. Action defines the self" (31; emphasis added). The creation of free communal spaces in Berlin apartments was always, in this way, a fraught process that drew on a fragile confluence of various identities and tactics. It should perhaps come as no surprise, therefore, that the solidarities and strains that punctuated communal life in the late 1960s mirrored the challenges that West German squatters would later face in the 1970s and 1980s. And yet, even as Kommune I disintegrated in a wave of anger and acrimony, other communal projects were founded in West Berlin that increasingly anticipated the occupation-based practices of squatters and other activists across West Germany.

It is with these very concerns in mind that the literary journal *Kursbuch* published a dossier on "Concrete Utopias" which called for the development of a "permanent process" of communalisation (Kirchknopf, 1968: 113). For members of Kommune I, there was, of course, no wider ecology of communal practices within the student movement and they claimed that no "preparatory structures" actually existed for "their way of living". "We are, in fact, the ringleaders and are indispensable," they insisted in a February 1968 article published in the radical underground magazine *Linkeck*.[18] As this chapter has argued, however, the figure of the commune spoke to a much wider and more intense field of political action. The former activist Peter Schneider described the mood in West Berlin at the time as an exhilarating atmosphere "that drove through the streets of Berlin [...] as an intoxicating wind [...] Everything seemed possible especially the impossible. We were carried along by this wind which called out to us to build a different society according to new rules. This was an intoxication without drugs, the intoxication of a 'historically necessary' and 'scientifically expressed' utopia that had taken possession of our hearts and minds" (Schneider, 2010: 13). Kommune I was, in this context, one of many experiments in shared communal living. This included Kommune 2 which occupied a 7-and-a-half-room apartment at Giesebrechtstraße 20 in Berlin-Charlottenburg in August 1967. Kommune 2 was largely made up of members of the SDS and took, if anything, a more rigorous approach to the collective organisation of everyday life from financial matters to the raising of children (see Kommune 2, 1969).[19] A number of other communal housing projects were also dotted across the city. The Linkeck-Kommune was formed in a small warehouse in Berlin-Neukölln in 1968 and lasted for only four months and later re-formed in Berlin-Schöneberg as the larger Bülow-Kommune. The Bülow-Kommune pushed the boundaries of communal living to extremes. All forms of privacy were vigorously proscribed and total openness was encouraged. If the commune was unsuccessful, the Anarsch-Kommune in Berlin-Charlottenburg adopted a more pragmatic and, in their eyes, anarchist approach to the negotiation of everyday tensions and differences. Theoretical disagreements were overcome, they argued, through the continuous adjustment of the commune's micropolitical dynamics (Knorr, 1969).

The expansion of communes across West Berlin (and the rest of West Germany) pointed to a further widening and deepening of the repertoire of contentious politics developed in the late 1960s by the extra-parliamentary opposition. In an important commentary published in the *Kursbuch* in 1969, Heide Berndt, the sociologist and former student of Theodor Adorno, highlighted the growing misconceptions surrounding the "political meaning of the term commune (*die politische Bedeutung des Begriffs Kommune*)". Berndt argued that this was a result of the close connection between a lack of public engagement and the unthinkingly neurotic behaviour of members of Kommune I. According to Berndt, approaches to communal living risked failure unless they actively sought to challenge the nature of human relationships. Berndt nevertheless advocated the further development of "concrete communes (*konkreter Kommunen*)" through which, she believed, the form and content of a radical future would be born (1969: 129, 144).

In the case of West Berlin, the founding of new communes increasingly spoke to the expanded field of goals and tactics that characterised the transformation of the New Left after the 'crisis year' of 1968 and the dissolution of the SDS (Brown, 2009: 6). This was a shift that saw the establishment of an autonomous women-only commune in Berlin-Steglitz in 1969 and the creation of a radical feminist infrastructure that encompassed bookshops, cafés, kindergartens, printing presses and women's centres (Amantine, 2011: 78; see Notz, 2006). It was also a shift marked, as Tim Brown has shown, by a "new focus on workers in general and young workers in particular" (2009: 7). A key feature of the new political landscape were the various Marxist-Leninist and/or Maoist cadre parties, the so-called *K-Gruppen*. These parties were accompanied by the emergence of 'rank and file groups' (*Basisgruppen*) as many young students sought to forge new solidarities and mobilise working-class youth (see Arps, 2011). In the end, these new groupings connected up with the more radical autonomous elements of the countercultural stream of the '68 movement. This included the scene around the West Berlin 'Blues' which combined a burgeoning youth culture based on rock 'n' roll and centred on clubs such as Club Ça Ira and Zodiak Free Arts Lab with an increasingly militant opposition to the state (see Brown, 2013: 70). These developments also depended on a "privileging of the local" as students and other activists directed their attention to the neighbourhood, the *Kiez*, as the drive to transform society shifted to concrete local struggles from youth politics to migrant housing, homelessness to urban regeneration (Brown, 2009: 7; see Bojadžijev and Perinelli, 2010).[20] The activities of the Wieland-Kommune and the Hash Rebel group that they spawned drew in, for the first time, a large group of local working-class youth from districts such as Kreuzberg who, in the wake of the Ohnesorg and Dutschke shootings, adopted a more confrontational approach that accepted the need for further escalation and the possibility of armed resistance. Whilst the communes that sprung up across West Berlin were spaces that were legally inhabited, the right to *occupy* space – legally or illegally – became a key tactic of protest as militant activists fought to preserve their right to spatial autonomy in the city.

The first space to be squatted was not actually in West Berlin. It was in the West German city of Cologne where, in 1970, a group of protesters occupied a block on Roßstraße in the working-class district of Ehrenfeld (Amantine, 2011: 12). The turn to occupation-based practices was, in many respects, a product of the dissolution of the German SDS in the autumn of 1969 out of which new connections and solidarities were developed between students, apprentices and other younger activists, a large number of whom had been in care or had run away from home. In Cologne, a support centre, the Social-Pedagogical Special Measures Cologne (*Sozialpädagogische Sondermassnahmen Köln* or SSK) was established as an alternative autonomous youth centre that fought neighbourhood gentrification while working with juveniles from youth homes and patients from psychiatric clinics (Sozialistische Selbsthilfe Köln, 1981; see also Brown, 2013; Gothe and Kippe, 1975; Kersting, 2006). It was forced to close in 1970, prompting a group of youth and their social workers to occupy the block on Roßstraße. Despite local support, the activists were soon evicted though a new occupation was initiated in an empty villa in Cologne-Marienburg a few months later. As the new occupants proclaimed, "we believe that the current situation, which is well-known, is unacceptable. In Cologne, there are over 1,700 people stuck vegetating in homeless shelters including 1,000 youth who have no home and are increasingly criminalised by the circumstances in which they are forced to live" (Pamphlet from occupation, quoted in Brandes and Schön, 1981: 41).

The next wave of occupations took place in Frankfurt in September 1970 as students linked up with homeless families and migrant workers to occupy an empty house at Eppsteiner Straße 47 in Frankfurt's Westend. The squatting of houses at Liebigstraße 20 and Cornelius Straße 24 followed a month later. The initial response to the occupations was positive and they were generally viewed in the press and by many politicians as a legitimate means through which to protest and challenge the abysmal conditions of housing in the city (Karakayali, 2000, 2005; see also Bojadžijev, 2008). The mood, however, changed as the rising number of occupations became part of a wider movement against housing speculation, rent hikes and unfettered urban redevelopment. "We will continue to occupy buildings," the squatters nevertheless insisted, "[and] declare war on a capitalist system that allows empty houses to remain empty (*um dem kapitalischen System den Kampf anzusagen, dass es ermöglicht, dass leere Häuser leerstehen*)" (quoted in Amantine, 2011: 15). As the protests and occupations escalated, new tactics were also adopted by many of the migrant workers involved in the wider housing movement. They turned, unsuccessfully, to rent strikes, a tactic they borrowed from Italy and radical groups such as *Lotta Continua* (The Struggle Continues) and *Unione Inquilini* (The Union of Tenants) who had become active in Frankfurt at the time.[21]

In the end, the solidarities formed between squatters and wider migrant communities were never really repeated elsewhere in West Germany and did not feature in the later history of squatting in the 1980s and 1990s (Karakayali, 2000). The main impulse for site occupation in the early 1970s thus remained the movement for independent and autonomous youth centres and the desire amongst young people to escape the unyielding predetermination of their lives and establish their own autonomous self-determined spaces.[22] It is against this backdrop that the first squat in Berlin began on 1 May 1970. While the occupation of a factory hall in the Märkishes Viertel was short-lived, other actions eventually followed. On 3 July 1971, over 300 students, activists and youth workers occupied two floors of an abandoned factory at 13 Mariannenplatz in the district of Kreuzberg with a view to creating a centre for disadvantaged and unemployed youth and "where we", as a pamphlet published by the activists declared, "can determine for ourselves what we do in our spare time".[23] Despite initial clashes with the police, municipal authorities eventually supported and legalised the initiative which included plans for a metal and wood workshop, a studio, a clinic and a theatre space.[24]

At the time, Kreuzberg was undergoing a process of urban renewal initiated by Mayor Willi Brandt in 1963. As I argued in Chapter 2, the squatting scene which first emerged in West Berlin in Kreuzberg in the early 1970s was a direct consequence of an endemic housing crisis that had its immediate origins in the vagaries of post-war reconstruction (Laurisch, 1981; Heyden and Schaber, 2008; MacDougall, 2011b). A key site in the politicisation of planning in Kreuzberg was the Bethanien Hospital complex at Mariannenplatz built between 1845 and 1847 and closed in 1970 though central heating and other utilities were kept on and the building was in "better condition than most homes in Kreuzberg".[25] The Bethanien had long been an important institution in the neighbourhood, having served a community that suffered from an overall lack of health and social services. Whilst the local *K-Gruppe* formed a neighbourhood (*Stadtteil*) committee and focused on the issue of housing as a means to connect with Kreuzberg's working-class inhabitants, it was the performative action repertoire described in this chapter that proved decisive in overcoming the impasse surrounding the status of the Bethanien. When activists occupied the abandoned factory at Mariannenplatz 13 in July 1971, they did so after a concert by the radical rock band Ton Steine Scherben who were themselves a product of late 1960s street theatre and included members from Hoffman's Comic Teater, the troupe that had played a key role in the occupation in the Märkisches Viertel a year earlier.[26] The concert at the Technical University of Berlin (Technische Universität Berlin, TU-Berlin) was organised by Peter Paul Zahl, editor of the leading Berlin radical organ *Agit 883* in conjunction with the Rote Hilfe (Brown, 2009: 11).[27] The concert was packed with students, young workers and activists from the increasingly militant scene in West Berlin. The lead singer of Ton Steine Scherben, Ralph Möbius (later known as Rio Reiser), implored the crowd to take action which led directly to the occupation at Mariannenstraße.

The relationship between performance and direct action took on an even more spectacular form in December 1971 after a teach-in at the Technical University to protest the shooting of the militant activist Georg von Rauch. The death of Rauch on 4 December 1971 in a shoot-out with police only served to heighten tensions between activists and the State and lent even greater urgency to the growing struggle over urban space in West Berlin. While the teach-in on the status of Bethanien complex had already been planned for 8 December, it was now accompanied by another concert by Ton Steine Scherben.[28] After the concert, a large group of youth including many involved in the creation of the *Kreuzberg Jugendzentrum* (Youth Centre) a few months earlier took the opportunity to squat in the abandoned Martha-Maria-Haus, a former residence for nuns and a part of the Bethanien Hospital complex.[29] The group that seized the house was once again made up of young workers and runaways as well as other groups: the local

Basis-Gruppe in Kreuzberg (the '*Basis-Gruppe Heim- und Lehrlingsarbeit*'), a workers collective (the '*Friedens – und Konfliktforschung*') and former drug users who had come together to form the local chapter of Release, a radical self-help initiative (Georg von Rauch-Haus, 1972: 9).[30] Members of the Berlin 'Blues' were also involved in the occupation. In a flyer circulated the next day, the squatters demanded (1) affordable housing for apprentices; (2) a new model of education and welfare for local working-class youth and runaways; and (3) a clinic and a counselling centre.[31]

The founding of the Georg von Rauch-Haus remains one of the defining moments within the broader history of squatting in West Berlin and was famously celebrated at the time by Ton Steine Scherben in the cult hit *Das Rauch-Haus Lied* with its rousing rhyming chorus: "*Ihr kriegt uns hier nicht raus – Das ist unser Haus*" ("You can't toss us out – This is our house") (Figure 3.1). A preliminary contract was signed with the Berlin Senate on 31 December 1971 and the house quickly became a key site within the radical scene in West Berlin (see Georg von

Figure 3.1   The Georg von Rauch-Haus (photo by author).

Rauch-Haus, 1972: 153–156). At the same time, differences between the various groups occupying the space led to a split between the apprentices and workers who occupied one floor and the runaways (*Trebegänger*) who turned the other floor into a project space that they shared with the *Basis-Gruppe* activists – mainly students – who worked with them. Tensions quickly escalated, although they were eventually resolved at a general assembly (*Plenum*) that brought together all the occupants. As a result, some residents chose to swap floors and a new communal kitchen was built in the basement (Georg von Rauch-Haus, 1972: 36–38). Negotiations also began between the occupants and local municipal authorities over the long-term use of the space. On 19 April 1972, the house was raided by the police looking for Michael "Bommi" Baumann, a former accomplice of Georg von Rauch-Haus and a member of the *2. Juni Bewegung* (June 2nd Movement). The police alleged that they recovered material that indicated that the house was being used as a 'bomb-making' factory (Brown, 2009: 12). Local CDU politicians attempted to use these allegations – what turned out to be five bottles of a cheap sparkling wine – to force the closure of the house. Negotiations with municipal authorities, including the Kreuzberg Councillor for Youth and Sport (*Stadtrat für Jugend und Sport*), Erwin Beck, nevertheless continued and eventually led to an extension of the existing lease in January 1973.[32] Notwithstanding changes to the original contract, the space has served as an autonomous housing project for more than 40 years.

The Georg von Rauch-Haus was one of many self-organised youth projects that emerged out of the first wave of squatting in West Germany in the early 1970s. In September 1972, a self-organised youth centre, The Drugstore, was established on Potsdamer Straße in Berlin-Schöneberg. A few months later, the occupants squatted a house at Wilhelmstraße 8 in Kreuzberg and were able to negotiate a contract for the formation of a self-organised collective (the *Sozialpädagogische Sondermaßnahmen Berlin*). The house became known as the Tommy Weisbecker-Haus after another militant activist and member of the Berlin Blues who was shot in 1972 by police in Augsburg. Similar projects were set up with varying degrees of success in Bielefeld, Dortmund, Hamburg and Hannover in the early 1970s. By the autumn of 1974, there were over 170 self-organised youth centres across West Germany. At the same time, squatters and other housing activists escalated their tactics in Frankfurt, occupying 20 separate houses across the city. The houses were forcibly cleared by the police culminating in a series of evictions on 21 February 1974. The demonstrations that ensued saw some of the worst street violence in West Germany since 1968.[33]

As the Frankfurt squatters themselves made clear in a special issue of the radical magazine, *Wir wollen alles*, it was the spectre of criminalisation and the problem of violence, more generally, that came to play an important role in the fracturing and splintering of the anti-authoritarian left in the early 1970s in West Germany (Figure 3.2).[34] While some activists embraced the militancy of groups such as the Red Army Faction and the June 2nd Movement, others increasingly

Figure 3.2   "The stones that hit your head are from the house you pulled down." Cover of *Sponti* magazine, *Wir wollen alles* (Nr. 13/14, 1974) documenting the Frankfurter *Häuserkampf* (personal collection, author).

looked to the formation of new oppositional geographies – both local and trans-national – as a source of renewed action and solidarity in the face of government countermeasures that affected the Left as a whole (Hanshew, 2012; Klimke, 2010; Tompkins, n.d.). However, for much of the 1970s, radical politics in West Germany was unable to resolve the problem of violence and what Jeremy Varon has called "the importance of being militant" (2004). As the violent actions of the RAF and other groups intensified, certain quarters of the New Left were overrun by a range of communist splinter groups that contributed to the further disorientation and fragmentation of the student movement and its offshoots. At the same time, and as I argue in the following chapter, the 1970s also bore witness to a significant reconstitution of oppositional practices which promoted the cultivation of activist spaces (bars, bookshops, cafés, kindergardens, music venues, workshops, etc.) and triggered, among other developments, a new expansive wave of squatting in Berlin and elsewhere in the country.

Spatialising the New Left?

This chapter makes a claim for the singular importance of thinking *geographically* about the history of the anti-authoritarian revolt in West Germany in the 1960s and 1970s. Whilst recent scholarship has begun to draw attention to the spatial dimensions of these protests, these accounts have tended to be descriptive and impressionistic (see especially Brown, 2013; Reichardt, 2014). In contrast, this chapter sets out to develop a critical geographical framework for rethinking the historical development of the New Left and the growing significance of West Berlin as a key site of protest. This is a framework, as the chapter shows, that focuses on the co-constitution of the urban and the political and the different ways in which the various practices adopted by students, activists and other citizens were *spatially generative*. The turn to squatting and site occupation as a legitimate tactic of protest came to represent, in this context, a further development in the repertoire of contention that emerged in the wake of the student movement in West Germany. In particular, it testified to "the rapid spread and mixture of new forms and tactics of protest that clearly distinguished the protest movements of '1968' from their historical predecessors" (Klimke and Scharloth, 2007: 5). These were, moreover, movements that depended, as recent scholarship has shown, on complex "geographies of connection" (see Featherstone, 2013) that linked activists in West Germany with struggles elsewhere in Europe and North America and militant protests across the global south (Klimke, 2010; Slobodian, 2012; Brown, 2013). As this chapter has argued, however, geography also mattered in other ways as public space – from city streets to courtrooms – became an important 'theatre of protest'. New forms of political contestation were, in other words, *spatial performances* that embodied a process of critical questioning forged out of a radical commitment to transform and refunction the nature of political action. And yet, for the 'revolution' to be *lived*, other transformations of space were still necessary (see Ross, 1988). It was ultimately, as this chapters shows, the figure of the commune that occupied a key role in linking political performance to the occupation of spaces where people came together to imagine a different form of collective life. The commune pointed, in other words, to the importance of assembling protest geographies where the "fulfilment of individual desires" was connected to a "political struggle against oppression and its various sources (*dem politischen Kampf gegen die Quellen der Unterdrückung*)" (Kommune 2, 1969: 9).

Protest and resistance thus depended on a new *spatial grammar* as early experiments in alternative forms of communal living in West Germany responded to the agitations of the late 1960s. A growing crackdown on public protest also served to exacerbate a shift in the spatialisation of protest and, by the early 1970s, activists had adopted increasingly radical strategies that included occupation and squatting. What some historians have described, in this context, as a retreat from the public sphere constituted, in fact, a new preoccupation with *Innerlichkeit* ('innerness') as activists looked 'inside' and turned to the emotional and material geographies of everyday life as a means of achieving autonomy and self-determination (Davis, 2008: 264; Häberlen and Smith, 2014). Intimate settings – squats, rural communes, youth centres, cafés, pubs, alternative presses, bookstores, youth centres, and parties – offered an expanded counter-geography for activists whilst hosting initiatives associated with closely-related campaigns and practices (antifascist organising, migrants' rights, feminist and queer politics, etc.). By the early 1980s, such an alternative milieu was responsible for the development of a radical infrastructure that encompassed a range of activities and sites whilst drawing in thousands of active participants and sympathisers across West Germany and in Berlin in particular (Reichardt, 2014: 13; see Huber, 1980). At the same time, a desire for a common 'form of life' also contributed to the growth of other local forms of activism such as the citizens' initiatives, or *Bürgerinitiativen*, that grew to prominence in the late 1970s and were to provide an organisational framework for the development of a second major wave of squatting in West Berlin (Markovits and Gorski, 1993). How, and to what extent, these practices came together in order to imagine and produce an alternative vision of the city will be the topic of the following chapters.

Notes

1   As the historian Aribert Reimann (2009: 123) has recently argued, the exact origins of the commune idea among activists in West Berlin in the 1960s remains contentious. Reimann suggests that the very first attempt to create a collective housing project dates back to 1964 and the Berlin art group Anschlag. Rudi Dutschke, as well as other SDS activists including Bernd Rabehl, were closely linked to the Anschlag group and it is clear that the idea of the commune as an alternative political space was already circulating within a wider activist community in the mid-1960s.

2   The SDS (the German Socialist Student League or the *Sozialistischer Deutscher Studentenbund*) was the most prominent leftist student organisation in West Germany in the first few decades after the World War II. Its origins can be traced back to the Association of Social Student Groups of Germany and Austria (1922–1929) and its successor, the Socialist Student Association (1929–1933). The SDS was founded on 3 September 1946 and was closely associated with the German Social Democratic Party in West Germany until disaffiliation in 1958.

3   In 1964, Rudi Dutschke and Bernd Rabehl set up an 'international working group' at the Free University in Berlin. The group included a number of Latin American, Haitian and Ethiopian students. It met regularly to discuss "the 'classics' as well as the newest texts of critical theory and Marxism" (Dutschke-Klotz, 1996: 58). Fanon's *The Wretched of the Earth* was presented to the group in December 1964 which was, in fact, many

months before the first German excerpt of the text was published in the journal *Kursbuch* (König, 1965). In the end, Dutschke was able to read the unpublished manuscript of the full German translation of the Fanon text due to his acquaintance with the translator, Traugott König (see Klimke, 2010: 266). Rabehl would later use the excerpted *Kursbuch* text in a seminar to new members of the SDS in the winter semester of 1965 (Slobodian, 2012: 59).

4   This was, in fact, the second meeting of student activists in Kochel. The first took place in December 1965 and also included Gretschen Dutschke-Klotz. The meeting was characterised by the tensions between Dutschke-Klotz and Kunzelmann who she accused of bullying and misogyny. Ulrich Enzensberger, a later member of Kommune I with Kunzelmann, has recently attempted to downplay the spat between Dutschke-Klotz and Kunzelmann and even went so far as to suggest that tensions were, in part, a result of Dutschke-Klotz's poor German (Enzensberger, 2004).

5   On Paul Tillich, see Rudi Dutschke, diary entry from 3.12.1964 (Dutschke-Klotz, 1996: 55).

6   *Der Abend*, 28.11.1966.

7   Hamburger Institute für Sozialforschung (hereafter HIS), Aktenbestand des Sozialistischen Anwaltskollektivs, File 56, testimony by Rainer Langhans on August 1, 1967.

8   The original text was written for a Subversive Aktion meeting that took place in Munich in April 1965.

9   Hans-Jürgen Krahl (1943–1970) was a student activist and member of the German SDS. Krahl studied with Theodor Adorno in Frankfurt. Alongside Rudi Dutschke, Krahl was a key theoretical figure within the German student movement. Krahl's most important text remains *Konstitution und Klassenkampf: Zur historiche Dialektik*.

10 Historians such as Martin Klimke and Maria Höhn have argued that the interactions between West German and American student activists played *the* definitive role in the adoption of direct action as a political strategy by German students and activists in the 1960s (Klimke, 2010; Höhn, 2002, 2008). While the significance of the American civil rights movement cannot be underplayed, recent scholarship has suggested that the emergence of new political movements in West Germany in the 1960s is in fact the product of a complex *transnational* set of exchanges and alliances that extended far beyond the United States (see especially Slobodian, 2012, 2013a, 2013b).

11 *Der Spiegel*, 17.4.1967, p. 17.

12 Freie Universität Berlin, "APO und Sozial Bewegungen" Archiv (hereafter APO-A), Flyers, Kommune I folder. All three flyers from Kommune I are dated 24 May 1967. The concluding words "burn, ware-house, burn!" were typed in English. According to Martin Klimke (2009: 265.n.10), the term "ware-house" was Dieter Kunzelmann's translation of the German word for department store (*Warenhaus*).

13 HIS, "Strafsache", Aktenbestand des Sozialistische Anwaltskollektivs, File 65, Kommune I, IIA Brandstiftung Diverses.

14 In May 2009, it was revealed that Kurras was in fact an agent of the Stasi and a member of the SED. In 2012, further research undertaken by Federal prosecutors and *Der Spiegel* showed that the shooting of Ohnesorg was not in self-defence as claimed by Kurras and that it was, if anything, premeditated. The report also implicated other police officers as evidence demonstrated that Kurras fired the shot apparently unchallenged from close range and surrounded by several police officers. Medical staff at

Moabit Hospital who carried out the post mortem on Ohnesorg were ordered to falsify their report (*Der Spiegel*, 22.1.2012, pp. 36–45; see Soukup, 2007).

15 *Der Spiegel*, 2.6.1997, p. 111.

16 In a 1969 column published in *Konkret*, Ulrike Meinhof noted that on 2 June the "realization that West German capital and the Iranian terror regime are closely allied was pounded into the students by the police. The same goes for the awareness that the opposition here – in the metropolitan centres – and the opposition in the Third World countries must work together" (Meinhof, 2008: 184).

17 Martin Klimke (2009) is right to point out that the relationship between the protest movements of the 1960s and 1970s and the West German legal system has largely been overlooked. Whilst a careful examination of the topic is beyond the compass of this book, legal geographies and their relationship to protest and dissent are explored in Chapters 4–6.

18 HIS, *Linkeck* 4 (1968), n.p.

19 Technically, Kommune 2 was established a few weeks before Kommune I in the headquarters of the SDS at Kürfurstendamm 140.

20 Interview with T.S. (Berlin, August 2010).

21 The rent strikes were ultimately unsuccessful as the various activist groupings in Frankfurt were unable to deal with the rising number of court cases which numbered 140 by 1973. Ninety percent were lost amidst growing tension between local activists and the wider migrant community. As Karakayali (2000, 2005) has argued, the paternalistic treatment of migrant activists only exacerbated the growing splits within the housing movement in Frankfurt.

22 Youth homes, as Timothy Brown (2013) has shown, were a key site of intervention for anti-authoritarian activists (265). This was perhaps best exemplified by a television film written by Ulrike Meinhof entitled *Bambule: Fürsorge – Sorge für wen?* which documented a riot in a girls' home in West Berlin. The film was pulled from the schedule a few days before it was meant to air in response to Meinhof's role in the freeing of Andreas Baader in May 1970 (see Meinhof, 1971).

23 Kreuzberg Museum, hereafter KM, untitled pamphlet, Ordner Hausbesetzungen.

24 *Berliner Zeitung*, 5.7.1971; *Der Tagesspiegel*, 6.7.1971.

25 APO-A, brochure, "Georg von Rauch-Haus: Ein Angriff das Georg v. Rauch Haus ist ein Anschlag gegen uns ALLE", p. 4, Rauch-Haus boxfile; original emphasis.

26 Ton Steine Scherben emerged in the late 1960s out of a radical street theatre tradition and the collaboration between three brothers (Gert, Peter and Ralph Möbius) whose group, Hoffmann's Comic Teater, performed on the streets and in youth homes across Berlin. The goal of the group was to develop a form of popular performance, drawing on everyday situations and conflicts, that would lead to direct political action. Ton Steine Scherben was formed as a follow-up project in Kreuzberg in August 1970 and was the first German rock group to sing exclusively in German. As Tim Brown (2009) points out, the turn to German was "meant to allow the group to connect as intimately as possible with its target audience, the apprentices and young workers of Kreuzberg" (8). In musical terms, the band adopted a sound that drew on English groups such as The Rolling Stones and The Kinks whilst anticipating the emergence of Punk in the late 1970s. The group eschewed the traditional system of production and distribution in favour of a DIY approach. They set up their own label to release their music through alternative channels tapping into a growing radical infrastructure

in West Berlin and West Germany. By the end of the 1970s, the band had sold over 300,000 copies of their music despite little advertising or radio airtime (Brown, 2009; see also Sichtermann et al. 2000).

27   The *Rote Hilfe* was the German affiliate of the International Red Aid and was originally active in Germany between 1921 and 1936. The organisation was founded to provide material and moral assistance to political prisoners. In the late 1960s and early 1970s, new unofficial groups that referred to themselves as *Rote Hilfe* began to appear across West Germany in response to statist repression. The *Rote Hilfe* e.V. was officially founded in 1975 and worked on behalf of groups that included the RAF.

28   Discussions between members of the Kreuzberg Jugendzentrum and local municipal authorities had already begun in the autumn of 1971. The occupants of the Georg von Rauch-Haus later alleged that they were promised one floor of the Martha-Maria-Haus during negotiations with local councilor Erwin Beck on 28 November 1971 (see Georg von Rauch-Haus, 1972: 9).

29   *Berliner Zeitung*, 9.12.71; *Der Tagesspiegel*, 9.12, 1971; APO-A, flyer, "Bethanien: Informationen aus dem Georg von Rauch-Haus" and "Wir machen was aus Bethanien," n.d. but likely December 1971, Rauch-Haus boxfile.

30   KM, flyer, untitled pamphlet, Ordner Hausbesetzungen. Release was an alternative health and welfare organisation set up in West Germany in 1969 by young people, many of whom were former drug users, in order to generate alternative solutions to the explosion of drug consumption in the late 1960s. The first Release organisation was founded in Hamburg at the end of 1969 and was modelled on a similar organisation that had been set up in London in 1967 (see Stephens, 2007: 186–188). The Berlin chapter of Release was located on Dennewitzstraße in Berlin-Schöneberg.

31   APO-A, flyer, "Gestern Abend. 600 Jugendliche besetzten das leere Bethanienn-Krankenhaus, 9.12.1971", Rauch-Haus boxfile.

32   Erwin Beck (1911–1988) was one of the few SPD politicians to actively support the German SDS and wider opposition movements in the late 1960s. Beck attempted to transform the Georg von Rauch-Haus into a wider social project that connected youth welfare with social work and local activism. The project received some support though was never realised. The majority of residents rejected, in particular, the permanent placement of social workers in the house as a form of compulsory social care.

33   See the special issue of the 'sponti' magazine *Wir wollen alles*. HIS, 13/14 (February/March, 1974), especially "Der Stein, den sie geschmissen haben", pp. 2–3; "Unsere Gewalt und ihre", pp. 4–7; "Spalten, kriminalisieren, draufschlagen," p. 9.

34   HIS, *Wir wollen alles*, 13/14 (February/March 1974). I am indebted to Matthew Hannah for this point.

Chapter Four
Antagonism and Repair

A total revolution – material, economic, social, political, psychic, cultural, erotic, etc. – seems to be in the offing, as though already immanent to the present. To change life, however, we must first change space.

Henri Lefebvre (1991: 190)

Come with us, said the donkey, we'll find something better than death everywhere.

Bremer Stadtmusikanten[1]

This chapter begins by recalling a trip to West Berlin made by the French activist, psychoanalyst and philosopher Félix Guattari at the end of January 1978. Guattari was speaking at the three-day TUNIX (Do Nothing) congress which had been organised by a range of leftist groups and was attended by Guattari, Franco "Bifo" Beradi, Michel Foucault, and Jean-Luc Godard among others (von Dirke, 1997; Hoffmann-Axthelm et al., 1979). The work of Guattari – himself a talismanic figure of the new social movements of the 1970s – had come to figure prominently in the evolution of the countercultural left in West Germany during the 1970s especially as more orthodox forms of extra-parliamentary opposition were challenged by Spontaneists (*Spontis*) and various anarchist groupings that identified themselves with a vibrant undogmatic Left and drew inspiration from other forms of political organisation and practice including, most notably, Italian workerism and its autonomist offshoots (see Brown, 2013; März, 2012). At the

Metropolitan Preoccupations: The Spatial Politics of Squatting in Berlin, First Edition.
Alexander Vasudevan.
© 2015 John Wiley & Sons, Ltd. Published 2015 by John Wiley & Sons, Ltd.

Treffen in TUNIX
WESTBERLIN 27.-29.1.78

KOMM MIT, SPRACH DER ESEL,
ETWAS BESSERES ALS DEN TOD
WERDEN WIR ÜBERALL FINDEN
(einer der Bremer Stadtmusikanten)

Figure 4.1   Cover of TUNIX conference flyer, December 1977/January 1978 (personal collection, author).

same time, the meeting of activists also took place under a cloud of leftist violence and statist repression as the struggle between the RAF and West German state agencies reached its bloody climax in the late summer and autumn of 1977. A second-generation of RAF activists attempted to force the West German government to release RAF members (Andreas Baader, Gudrun Ensslin, Jan-Carl Raspe) still held in custody. The Chief Federal Prosecutor, Siegfried Buback, was assassinated, as was the director of the Dresdner Bank, Jürgen Ponto. This was followed by the abduction of leading industrialist and head of the German Business Association, Hanns-Martin Schleyer, and the hijacking of a Lufthansa passenger jet by Palestinian insurgents in a coordinated operation. On 17 October 1977, West German security forces successfully stormed the Lufthansa plane at an airfield in Mogadishu, Somalia, freeing all the passengers. The next morning, Baader, Ensslin and Raspe were all found dead in their cells at the Stuttgart-Stammheim prison in what appeared to be suicides. In retaliation, Schleyer was shot by his RAF captors and his body left in the trunk of a car near the Belgian border (Hannah, 2012: 119; see Aust, 1985; Aust and Winkler, 2008).

The events of the German Autumn (*Deutsche Herbst*) are traditionally under-stood as marking the end of the anti-authoritarian struggle in West Germany. They also coincided, as a number of scholars have recently suggested, with the deployment of exceptional security measures (stop-and-search actions, emergency legal decrees, employment bans, press censorship) that prompted many young West Germans in particular to believe that they were now living in a 'security society' governed and policed by a 'security state' (Hannah, 2012: 123; see also Hannah, 2010; März, 2012).[2] Against this new and growing politics of repression, it is perhaps not surprising, as Sabine von Dirke has pointed out, that those attending the TUNIX meeting in January 1978 sought to remedy the 'identity crisis' that the anti-authoritarian Left had suffered in the context of escalating violence and the government's sweeping punitive countermeasures. If activists were keen to express their "dissatisfaction with the sociopolitical situation in the Federal Republic" they were equally concerned with rethinking the "left's tradi-tional strategies of opposition" and they hoped that the meeting would provide a "forum for self-assertion and mutual encouragement on the part of the alternative culture" (von Dirke, 1997: 111–112).

In this way, the TUNIX meeting returned to an enduring, and indeed unre-solved, tactical question that underpinned the development of the extra-parliamentary opposition in West Germany: namely, whether radical social change could be achieved through the takeover of existing structures of political and economic power or whether it was better to assemble "oppositional geographies" that would serve as a blueprint for an alternative social order (Hannah, 2010: 18). The questions originally posed by the protagonists of 1968 – What constitutes a revolution? Is it possible to reconcile personal liberation with social transforma-tion? How do we connect the possibilities of daily life with wider structures of protest and revolt? – were the same questions facing the activists that met in West Berlin in 1978 (see Brown, 2013: 371). Guattari's own address at the meeting seized on these concerns (Figure 4.2). Whilst Guattari drew attention to the brutal repression practised by the state, he also described a "softer, yet also more systematic and deceptive repression [that came to be] ... diffused into all the pores of society". Echoing the work of Michel Foucault with whom he shared the stage in Berlin, the main animus behind Guattari's address was to challenge what he saw as a new alignment of law, security and biopower.[3] To counter these developments, Guattari advocated a *micropolitcs* that promised "new spaces of liberty" and the composition of transversal alliances and solidarities that would connect radical groups and organisations across Europe and support the release of "new perspec-tives on common struggle". At stake here for Guattari, was the necessary time and space for creating an alternative infrastructure to support and sustain practices of reciprocity, sociability and world-building that moved beyond conventional for-mulations or bracketings of the 'political' (see Vasudevan, 2011a). The possibility for social change, according to Guattari, was to be found therefore less in tradi-tional politics than in a kind of 'molecular revolution' which demanded new modes

Figure 4.2   The TUNIX conference in West Berlin, January 1978 (Klaus Mehner, BerlinPressServices).

of connection and new "existential territories" to borrow his own useful phrasing (Guattari, 2009 [1978]: 98, 99, 96).

Whilst it is admittedly impossible to measure the immediate impact of Guattari's address, it is worth pausing to acknowledge the direct contribution that the TUNIX meeting made to the recomposition of the alternative scene in West Germany and elsewhere. The event attracted an estimated 20,000 attendees from around the world, far surpassing the expectations of its organisers (Brown, 2013: 355). It also brought together the various strands of the undogmatic left in Europe who all shared a desire for a "re-envisioning of the nature of politics and the terms in which it was carried out" (Brown, 2013: 355). As much as the TUNIX meeting thus served as a launching point for the alternative movements of the 1970s and 1980s (Brown, 2013: 362), the development of what Guattari described as "existential territories" found its most concrete expression in a major new wave of squatting in cities such as West Berlin and, later, Hamburg. The illegal occupation of empty flats and buildings was itself hardly new. As I show in this chapter, the large-scale occupation of flats in West Berlin marked, if anything, a further development in the history of radical urban politics in West Germany rather than a complete break with the repertoire of contentious politics that had consolidated themselves in the preceding decade. The chapter challenges recent accounts of the TUNIX meeting and its aftermath. It argues that the meeting did not represent, as the historian Timothy Brown (2013) has recently argued, an end to the anti-authoritarian revolt in West Germany (362). If many activists including Guattari (1986) spoke of the late 1970s and early 1980s as "the winter years", the defeat and retreat of radical movements across Europe and the existential fallout from this withdrawal also prompted a renewed commitment to the cultivation of alternative spaces of subversion, solidarity and self-determination (see Negri, 2013). I therefore seek to append a sympathetic critique to Brown's recent association of the TUNIX meeting with the eclipse of the extra-parliamentary opposition in West Germany. My main aim here is to show how the city of West Berlin remained a key site of struggle in what had become known as the *Häuserkampf* ('the housing struggle') and that the emergence of urban squatting as a radical social movement in the city depended on the *extension* of occupation-based tactics first developed in the late 1960s and early 1970s. This was, after all, a struggle over the right to the city and against the violent predations of capitalist accumulation. It was also a struggle that sought to *connect* a housing crisis shaped by the logics of creative destruction and accumulation by dispossession with a crisis of dwelling and a desire shared by many to carve out sites of autonomy and emancipation.

At the heart of the chapter is a detailed reconstruction of the period between 1979 and 1984 which represented the high point for the squatting movement in West Berlin. If Chapter 3 established the importance of 'occupation' to the repertoire of contention mobilised and performed by the New Left in West Germany, this chapter further examines the practice of squatting as an act of collective *world-making* through which an alternative understanding of city life was (quite literally) built. It concentrates, in particular, on how squatted spaces were assembled and sustained on an everyday basis. The chapter shows how this depended in no small part on a politics of adaptation, mending and repair that served as a direct response to an endemic housing crisis characterised by top-down planning initiatives, rampant property speculation and local corruption. Squatters in West Berlin often confronted abandoned spaces that required significant renovation, and in this chapter I offer a thick description of the wide range of practices and tactics deployed by squatters as they challenged and were later compelled, in many cases, to accommodate existing property regimes. Whilst my main aim is to examine the everyday practices of squatters as a radical *makeshift urbanism*, I also seek to consider the complex constellation of affects, emotions and feelings and the decisive role that they came to play in the social life of a squatted house (see Brown and Pickerill, 2009; Gould, 2009). According to this view, the activities of squatters embodied a form of *emotional labour* through which the boundaries of 'activism' and 'the political' were constantly made, unmade and remade.

The chapter is divided into four parts. The first part of the chapter focuses on the evolution of the radical scene in Berlin in the immediate lead-up to the TUNIX Congress in January 1978. The remainder of the chapter examines the consequential emergence of the squatting movement in West Berlin in three stages. The first zooms in on the key period between 1979 and the 'hot summer' of 1981 at which point over 169 houses were occupied across West Berlin. The second highlights the material and emotional geographies produced by squatters in the city and how they served as an active archive of practices that promoted the development of new forms of collective living. The final focuses on the period after 1981 and the complex transformation and dissolution of the squatting scene through protracted negotiations with West Berlin authorities, the legalisation of some occupied houses and the 'pacification' of the more militant elements of the movement through criminalisation and eviction (see Katz and Mayer, 1985; Holm and Kuhn, 2010). Taken together, the chapter demonstrates how the everyday spatial practices of squatters in West Berlin represented, for some, an act of militant antagonism and insurgency and, for others, a trade-off between existing political institutions and forms of citizen self-help and self-management. At the same time, the chapter also offers an opportunity to examine how activists responded to decline and failure, dissent and violence. Squatted spaces, it concludes, were both sites of liberation and possibility and sources of intense conflict and struggle.

"Wir wollen alles und wollen es jetzt": Re-imagining Protest in 1970s Berlin

The immediate origins of the TUNIX congress can be traced to the mid-1970s and the Tiergarten Park in Berlin where, every Saturday, a group of students met to play football on a field behind the Berlin Kongreshalle (see März, 2012).[4] The students from the Free University were members of the Sponti scene – a loose grouping of activists and student radicals influenced by anarchism who favoured an 'undogmatic' approach to political action (Brown, 2013: 83–84; see Geronimo, 1992). The Sponti scene was one of many responses to a crisis in the extra-parliamentary opposition in the late 1960s which had become the source of an intensive debate about organisation, strategy and future direction. Many activists formed *K-Gruppen*, groups modelled on orthodox Marxist-Leninist lines, that "returned evermore single-mindedly to the words and deeds of the great men of the socialist-revolutionary tradition" (Brown, 2013: 109). The Spontis, on the other hand, advocated an anti-authoritarian form of politics that privileged spontaneity and contingency in place of certainty and orthodoxy. Unlike the *K-Gruppen*, they adopted an alternative course of revolution which did not see the working-class as simply a monolithic entity waiting for a vanguard to lead them but rather as an autonomous revolutionary subject in its own right. They drew inspiration, in this context, from their counterparts in France and especially Italy where groups such as Workers' Autonomy (*Autonomia Operaio*), Workers' Power (*Potere Operaio*) and The Struggle Continues (*Lotta Continua*) became increasingly preoccupied with the "emerging autonomy of the working class with respect to capital, that is, its power to generate and sustain social forms and structures of value independent of capitalist relations of production" (Hardt, 1996: 2; see Geronimo, 1992: 19–22; Wright, 2002). At stake here, as argued by Mario Tronti, one of Italian *Operaismo*'s early interlocutors, was a form of working-class *self-organisation* that adopted a "wholly alternative content" which refused to "function as an articulation of capitalist society" and to therefore "act as an active partner in the whole social process". By the early 1970s, such a "strategy of refusal" had prompted the struggle in Italy to shift focus from the factory to the city and an expanded *geography of protest* that encompassed workplace occupations, pirate radio stations and countless squats (Tronti, 1965).

Like Italian autonomism and the workerist tradition it evolved out of, the German Sponti scene was committed to a "total, anticapitalist – and militant – struggle against the complete domination of capital over the sphere of production and reproduction" (Della Porta, 2006: 103). The Spontis eschewed the strict hierarchy and organisation that characterised the *K-Gruppen*, preferring to form loosely-organised collectives that often worked in volatile communities within schools, local neighbourhoods, and the workplace. Sponti activists played a crucial role in the formation of 'Factory Project Groups' (*Betriebsprojektgruppen*) that attempted to organise with and support immigrant guest workers, many of whom were Italian (Karakayali, 2000, 2005; see also Bojadžijev, 2008). These groups scored some notable local successes, such as the October 1971 action at the Opel factory in Rüsselheim and the August 1973 strike at the Ford factory in Cologne (Brown, 2013: 274; Arps, 2011). The Spontis sought to translate these minor victories into other campaigns, including the struggle for housing in Frankfurt described in Chapter 3 and which ultimately gained considerable popular support through an action repertoire that included "countless demonstrations, street battles, rent strikes and building seizures" (Brown, 2013: 275; Geronimo, 1992: 22). At the same time, the Sponti scene was also responsible for a number of underground publications, including magazines such as *Schwarze Protokolle, Carlo Sponti, Wir wollen alles* and *InfoBUG* (*Info Berliner Undogmatische Gruppen*). With a wide circulation across West Germany, it was *InfoBUG*, in particular, which became the critical forum for the discussion of issues affecting leftists in the 1970s. From the spring of 1975 onwards, issues of the magazine were routinely confiscated by the state. In February 1977, legal action was initiated against the publishers, the Agitdruck collective. A few months later, four members of the collective were arrested on charges of supporting terrorism (see Brown, 2013: 353–354).

The emergence of the Sponti scene in West Germany testified to a significant expansion in the scale and reach of leftist activism during the 1970s (Koenen, 2001; Reichardt and Siegfried, 2010; Reichardt, 2014). It also spoke to a growing dissatisfaction with the escalating militancy of groups such as the RAF and the Tupamaros who had all but jettisoned their political activities for an armed struggle that focused on securing the freedom of their fellow comrades in jail (März, 2012: 204). If the upsurge of violence, which reached its apogee in the events of the German Autumn, has tended to dominate the historiography of the new Left in West Germany, a weariness with the RAF and its tactics prompted many Spontis to search out and cultivate new political possibilities that have so far received far less attention. As one former activist noted, "the whole direction was wrong. We needed to change course". "Somewhere between Stammheim and Mogadishu," he continued, "our dreams were extinguished." He was supported by a former comrade who recalled how "things had spiralled completely out of control", while another activist reflected on the "need to be clear that we must avoid violence".[5]

The question of violence did not, however, go away and remained one of the basic antinomies that shaped the development of the New Left in West Germany in the late 1970s and early 1980s. Even the operational imperatives that underpinned the Sponti scene – autonomous, anti-authoritarian, self-organisational – offered no unified position. Nevertheless, the Spontis did turn their back on the RAF in search of new forms of action and solidarity. The TUNIX Congress was modelled, in this context, on an international conference that took place in September 1977 in Bologna Italy, the 'Congress against the Repression in Italy' (*il Convegno contro la repressione*). At least 70,000 participants attended the Congress which represented a defining moment in the history of Italian *Autonomia*, an intense period of political, social and cultural experimentation that reached its high point in Italy in the summer of 1977 and that was ultimately the subject of a brutal crackdown by the state (Wright, 2002). The defeat and political repression should not, however, detract from the historical and geographical significance of Italian *Autonomia*. As Félix Guattari noted in a 1982 interview with Sylvère Lotringer, "that the Italian *Autonomia* was wiped out proves nothing at all. From time to time, a kind of social chemistry provides us with a glimpse of what could be another type of organization" (Guattari and Lotringer, 2009 [1982]: 119).

Between September 23 and 25, the university town of Bologna was transformed into a radical festival that, for many West German activists in particular, offered just such a glimpse into a different political future. The writer Ronald Glomb described the intense carnivalesque atmosphere and the blurring of boundaries between the political and the cultural, between protest and performance. The various attendees, made "their own circus out of life," he observed, "singing, dancing, yelling, fucking, fighting (*die aus dem Leben, ihren eigenen Zirkus machen, singend, tanzend, brüllend, bumsend, kämpfend*)" (Glomb, 1979: 139).[6] Amidst the wide range of groups that attended the meeting, it was the creative actions of the Italian 'Urban Indians' (*Indiani Metropolitani*) – from wall murals to street theatre – that offered perhaps the most appealing alternative vision to the violent impasse that characterised the scene back in West Germany.[7] The 'Urban Indians' were often seen as the most radical and creative group within the autonomous movement in Italy adopting a political programme that focused, as one contemporary commentator reported, on a "politics of free spaces

(*Politik der Freiräume*)" through which the needs of everyday life were politicised and experienced as collective forms of self-determination (Geronimo, 1992: 45). The latest in a string of transnational personae adopted by leftist activists in West Germany, the figure of the 'Urban Indian' became extremely influential in the lead-up to the TUNIX Congress though some, including the political scientist Johannes Agnoli, warned against a crude appropriation of a movement that failed to fully acknowledge the content and context of the politics at stake (Agnoli, 1979: 88; see Slobodian, 2012).

Whatever the case, the planning for the TUNIX Congress was firmly embedded within wider, often unruly, geographies of political engagement that depended on forms of worldliness, mobility and connection that were resolutely translocal (Featherstone, 2013: 1408). Recent scholarship by geographers has drawn attention to the relationalities generated through subaltern political activity and the significance of such an approach to debates on the co-constitution of space and politics (Featherstone, 2013: 417; see also Featherstone, 2012). While this work challenges the association of radical forms of political action with "bounded or restricted localised spatialities", it was, if anything, a sense of isolation and boundedness that increasingly shaped the geographical imagination of many militant activists in West Germany in the 1970s. This was an argument developed to great effect by Peter Brückner, the prominent professor of psychiatry who would later share the podium with Michel Foucault and Félix Guattari at TUNIX.[8] In a 1977 book, Brückner, a key figure in the West German anti-psychiatry movement, explored the growing divergence between mainstream and alternative cultures (Hannah, 2012: 130). Through a close psychoanalytical reading of the writings of Ulrike Meinhof, Brückner accounted for the gradual dissolution of Meinhof's sense of proportion as a product of her isolation from social life which led, so he argued, to a "disappearance of place and time". "This disappearance of place and time," he argued, "as though imperialism had really swallowed up the life-space (*Lebensgelände*) of social processes (and with it the 'places,' the topography of everyday being-in-the-world), [had] torn the people from the temporality of their lives, is expressed and dealt with by Ulrike M. Meinhof—and by other comrades—as an essential and communicable inner experience…" (Brückner, [2001] 1977: 172).

Meinhof's unique ability to communicate this experience in her writing may have convinced some that this was, in fact, a more than adequate description of the circumstances facing the militant left in West Germany. For many in the Sponti scene, however, this was an experience that foreclosed any meaningful attempt to generate other forms of shared political action. It is hardly surprising, therefore, that the "impulses fueling TUNIX were transnational in orientation and genesis" (Brown, 2013: 358). And yet, they were also stubbornly local and connected to activist infrastructures and 'scenes' while speaking to issues that were rooted in neighbourhood struggles in West Berlin and elsewhere. The group of footballer-activists who met in the Tiergarten every week, including Johannes Eisenberg,

Figure 4.3   Programme for TUNIX conference, *InfoBUG*, December 1977/January 1978 (personal collection, author).

Peter Hilleband, Stefan König, Diethard Küster and Harald Pfeffer, were also linked to a network of pub collectives in West Berlin and to two alternative book-stores that were located in Berlin-Charlottenburg where they met Renée Zucker and Monika Döring (März, 2012: 206).[9] All would later play a prominent role in the organisation of the TUNIX congress.

The idea for the congress was finally launched in December 1977 in the Ambrosius pub after a Saturday afternoon football match. As Küstner, one of the organisers recalled, "it was meant to be a farewell, an intoxicating festival of utopians (*ein rauschendes Fest der Utopien*)". "The name TUNIX," he continued, "expressed the life feeling of many students, for whom work was a synonym for the deprivation of freedom."[10] For the planners, the meeting was, of course, far more than a simple farewell. It represented an opportunity to extend and revitalise the New Left in West Germany. According to a programme flyer circulated in advance of the meeting (Figure 4.3),

> For years we have believed that things could be changed by actions under the motto 'away with…' and 'down with…', if only one tried at it hard enough. Our fantasy was mangled, euthanized or shattered (*Unsere Phantasie wurde darüber verstümmelt,*

eingeschläfert oder verschüttet). Instead of engaging at this traditional level of resis-tance as we have always done, this time we want not only to talk about new forms of resistance, but practice them in the course of our gathering. We want to develop new ideas for a new struggle, that we ourselves decide upon, and not let ourselves be dic-tated to by the technicians of 'Modell Deutchland' (*Wir wollen neue Ideen für einen neuen Kampf entwickeln, den wir selbst bestimmen und uns nicht von den Technikern des 'Modell Deutschland'*).[11]

The Congress was thus imagined as a 'trip' - a line of flight - out of a West Germany dominated by a repressive market-driven 'performance society' that was best captured, for many organisers, in the slogan used by the SPD in the 1976 election, '*Modell Deutschland*'. "We no longer want to do the same job, see the same faces over and over again," proclaimed the congress announcement, "they have bossed us around enough, have censored our thoughts and ideas, checked our apartments and passports, and bashed us in the face. We will no longer let ourselves be bottled up, made small, and made the same (*Wir lassen uns nicht mehr einmachen und kleinmachen und gleichmachen*)." The announcement also made its debt to Italian *Autonomia* abundantly clear: "We want the MOST FOR EVERYONE! Everyone can paint, sing and formulate their own ideas and slogans though we will nevertheless – and perhaps therefore – fight together. WE WANT EVERYTHING AND WE WANT IT NOW!!!!!!."[12]

The TUNIX organisers took advantage of West Berlin's large network of independent DIY printers to produce and distribute invitations in December across West Germany and elsewhere in Europe. By the time the organisers had returned from the holidays, thousands – to their astonishment – had signed up. In addition, a number of prominent speakers had also been confirmed, including the philosophers Michel Foucault and Félix Guattari, and the lawyer and associate of the Socialist Lawyer's Collective (*Sozialistichen Anwaltskollektivs*), Hans-Christian Ströbele. The backroom of a bookstore at Savignyplatz 5 served as the unofficial headquarters of the congress as requests to participate poured in from all corners of the extra-parliamentary left. The organisers booked the Technische Universität Berlin (TU-Berlin) with the assistance and support of the SPD senator Peter Glotz. Glotz agreed, in turn, to participate in a podium discussion with Peter Brückner, Daniel Cohn-Bendit, the President of the Free University, Eberhard Lämmert, and other representatives from the Sponti Scene on the growing divisions between mainstream and alternative culture (Brown, 2013: 357).

The congress took place from 27–29 January 1978. The programme included panels on 'autonomous queer theory', 'feminism and ecology', 'knowledge-utopia-resistance', 'alternative media practice', 'radical pub collectives' and a host of sessions on environmental issues. A major keynote session on 'psychiatry-antipsychiatry' was held in the main TU auditorium and featured Brückner, Foucault and Guattari. Pressing contemporary issues from the 'Ban on Careers' (*Berufsverbot*) to the imprisonment of the Agitdruck printers were also represented.

The organisers scheduled a wide range of cultural events – agitprop cabaret, street theatre, music gigs and puppet shows – which took place between panel sessions. A two-day TUNIX film festival was held in conjunction with the congress and showcased the latest work from underground and more established directors. At the same time, a number of panel sessions were organised that spoke to more modest local geographies and the neighbourhood struggles that they encompassed: a panel on autonomous youth centres featuring speakers from the Georg von Rauch-Haus in Kreuzberg; a series of sessions on city district work (*Stadtteilarbeit*) that featured activists who were involved in the squatting of an abandoned fire station in the same district in May 1977; and, finally, a standalone session on the struggle against uneven urban development.[13]

Despite criticisms from some quarters, the TUNIX congress was widely seen as a success.[14] While it is, of course, impossible to gauge the full contribution that the TUNIX meeting made to the recomposition of the radical leftist scene in West Germany, it was certainly responsible for the launching of a number of important practical initiatives from the founding of the left-leaning *die tageszeitung* newspaper to the development of the German Green Party (Brown, 2013: 361). The gathering's emotional undercurrents were equally significant as the French philosopher Guy Hocquenghem argued in an article in the *Pflasterstrand*, the magazine of the Frankfurt Sponti scene. For Hocquenghem, the TUNIX meeting represented an *affective* release from the "militant nervousness" of the German Autumn. It offered, he argued, breathing space for those groups who "did not wish to be crushed between the state and terrorism".[15] The freeing-up of what Hocquenghem described as "spaces of otherness" also provided, in his view, an important context for the exploration of new protest techniques that built on and extended the action repertoire developed during the 1960s and 1970s and to which the active assembling of alternative spaces had already assumed a certain significance (Reichardt, 2014). It should come as no surprise, therefore, that plans for TUNIX coincided with the founding of a new citizen's initiative and alternative newspaper in Kreuzberg in late 1977 by a group of activists who, like their TUNIX counterparts, met weekly in a local pub to discuss and organise against the eradication of the things that they "love[d] about Kreuzberg – the small shops, the pubs, the backyard workshops".[16] The group was first known as *Stammtisch SO36* after the district's postcode and the informal setting in which they met. They soon changed their name to BI SO36 which was short for *Bürgerinitiative SO36* (or Citizen's Initiative SO36). The group was responsible for the publication of the *Südost Express*, an alternative '*Kiez*' or neighbourhood magazine that first appeared in December 1977. The magazine quickly became a key source of information in Kreuzberg on a range of housing-related issues, including tenant rights, rising rents, local corruption, construction scandals and urban 'regeneration' (see Laurisch, 1981). It was the various protest actions orchestrated by the group, however, that ultimately led to the first major wave of squatting in West Berlin.

From Demolition to Occupation: Squatting the City

In Chapter 3, it was argued that occupation-based practices, including squatting, represented an important example of the new forms and tactics of protest that characterised the emergence of the extra-parliamentary opposition in West Germany in the late 1960s and early 1970s. The chapter examined how these practices *endured* and produced new sites of opposition, resistance and autonomy that re-imagined the city as a site of "political action and revolt" (Harvey, 2012: 118–119). Whilst West Berlin undoubtedly occupied a privileged place in a wider activist imaginary, the early history of squatting in West Germany encompassed an expansive geography that included well-developed scenes in other cities such as Frankfurt, Hamburg and Freiburg (Amantine, 2012; Dellwo and Baer, 2012, 2013). It was, in fact, in Frankfurt where the so-called 'struggle for housing' or *Häuserkampf* first became a "signature radical campaign of the 1970s" (Brown, 2013: 273). The widespread occupation of apartment buildings in the early 1970s even succeeded for a brief time in transforming parts of the city's Westend into a self-determined autonomous zone (see Geronimo, 1992; Dellwo and Baer, 2012).

In West Berlin, the first squatted spaces in the district of Kreuzberg were closely connected to the movement for autonomous youth centres and the desire amongst many young people for freedom and self-determination.[17] Iconic sites such as the Georg von Rauch-Haus and the Tommy Weisbecker-Haus spoke to the needs of many working-class youth and quickly became major 'scene' locations for young runaways, drug users and radicals from a variety of political groupings (Brown, 2009). They were also sites embedded within a wider constellation of struggles over social inequality, housing precarity and urban regeneration. These struggles were hardly new and, as I argued in Chapter 2, the actions of squatters in the 1970s and 1980s in West Berlin were, in fact, part of a complex history of resistance to capitalist urbanisation in Berlin and the further extension of long-standing critiques of creative destruction and modernist urban renewal (see MacDougall, 2011a). To this end, I reconstructed a genealogy of these critiques that can be traced back to the 1960s and growing opposition from planners, architects and community organisers to top-down planning as an idea and practice. At the same time, I uncovered an even older sedimented historical geography of dissent over housing in Berlin from the barelife urbanism of squatter settlements in the 1860s and 1870s to widespread strikes over rising rents in the 1920s and 1930s (Kowalczuk, 1992). In the remainder of this chapter, I seek to build on the arguments outlined in Chapters 2 and 3 respectively and shift registers, zooming in on the everyday tactics adopted by squatters that emerged alongside the professional critiques of the 1960s and 1970s. Whilst it was planners who first rediscovered the value of Berlin's historical housing stock in neighbourhoods such as Kreuzberg, it was ultimately squatters and other housing activists who *re-animated* these spaces as sites of autonomy, resistance and self-determination (see Vasudevan, 2011a; McFarlane and Vasudevan, 2013).

By the time of the TUNIX conference in 1978, the district of Kreuzberg in Berlin had already acquired an almost mythical status in the city and West Germany more widely as the "symbolic site of political, cultural and ethnic difference in Germany". The neighbourhood was often framed as a 'ghetto' populated by anarchists, bohemians, dropouts, punks, intellectuals, students and migrant workers from Turkey (see MacDougall, 2011a: 125; see Lang, 1998). The neighbourhood's self-image was further shaped by the presence of the Berlin Wall which transformed Kreuzberg into a "depopulated cul-de-sac" shaped by falling housing prices and a prodevelopment lobby preoccupied with "shifting margins of profitability and revalorization" (Sheridan, 2007: 101; Blomley, 2004: 79). Semi-derelict housing stock from the 19th century, abandoned factory spaces and vacant tracts of land thus remained under-developed while low-income residents struggled to find affordable housing. Kreuzberg represented, so it seemed, yet another example of the new form of urban marginality that increasingly shaped advanced capitalist societies in the early 1970s (Wacquant, 2007).

As I suggested in Chapter 2, the first urban renewal programme launched by the West Berlin Senate in the 1960s had come under sharp criticism, and by the mid-1970s, it was clear to critics – from planners and architects to grassroots neighbourhood organisations – that regeneration merely served as a pretext for higher rents, displacement and widespread demolition. Many residents living in parts of Kreuzberg marked for inner-city renewal had already received eviction notices and had been 'decanted' to newly-built satellite cities on the outskirts of West Berlin. Urban renewal agents favoured, in turn, the demolition of older housing stock as they were able to apply for major federal subsidies and tax exemptions to support new build developments (see Tomann, 1996).[18] Combined with the economic recession of the early 1970s, whole neighbourhood blocks in Kreuzberg fell into serious disrepair as they 'waited' for the city to implement renewal plans. Owners had no incentive to undertake even the most basic of renovations and this intentional neglect only served to further diminish the quality of the housing stock such that demolition remained the only option while assuring owners generous pecuniary benefits (MacDougall, 2011a: 89).

In the end, it was a statement issued in 1973 by the local district board of the Protestant Church (*Evangelische Kirche*) that seemed to capture the full impact of urban redevelopment in Kreuzberg. "Urban renewal," it stated, "isn't living up to its promise of a planning practice that will provide Kreuzberg residents [...] with a better quality of life. Instead, it only benefits the capitalist interests of housing associations and private owners and serves as a playground for planners and architects" (Evangelische in Berlin-Kreuzberg, 1973: n.p.). Many of those planners and architects came, in fact, to a similar conclusion and argued against demolition and for the preservation of the neighbourhood's old tenement blocks (*Mietskasernen*). Facing growing dissent, local politicians promoted a shift to greater citizen engagement. They endorsed and formalised "The Strategies for Kreuzberg" competition that was launched on 3 March 1977 as a strategic partnership between the Berlin Senate for Building and Construction and the Protestant Church of Berlin-Brandenburg (MacDougall, 2011b: 166). It placed particular emphasis on the co-production and co-management of planning principles and placed local residents, in theory at least, at the heart of the decision-making process (see Strategien für Kreuzberg, 1978).

In practice, however, the city's commitment to resident participation was lukewarm at best. At the very same time that the city announced the "Strategies" competition, bulldozers appeared at a 19th-century fire hall and pumping station on Reichenberger Straße in anticipation of the building's demolition to make way for a new school and day-care centre. The move antagonised local Kreuzberg residents and activists who interpreted the action as a continuation of longstanding policies that undermined and threatened the neighbourhood's existing network of informal cultural and economic ties. These ties were a product of the many small factories and workshops that still co-existed alongside older blocks of housing. What was widely known as the 'Kreuzberg Mix' (*Kreuzberg Mischung*) blurred the boundaries between work and leisure while producing an everyday geography shaped by deeply sedimented patterns of working-class sociability that, in turn, encompassed local corner stores, pubs, and mixed-used courtyards (see Niethammer and Bruggermeier 1976; von Saldern, 1995). These spaces functioned as both a social network through which information on jobs and housing was circulated and a base for political mobilisation (Beier, 1982: 264).

It is therefore not surprising that local residents responded to plans to demolish the old fire hall by setting up a citizen's initiative (*Bürgerinitiative Feuerwache*) that highlighted the needs and wishes of the neighbourhood. The initiative promoted an alternative model that argued for the preservation of the station as a community social centre (*Stadtteilzentrum*) that would host a range of different groups (a women's group, a community crisis centre, addiction counsellors, etc.). On 5 May 1977, members of the initiative squatted the Reichenberger fire hall proclaiming that they shared a "conviction that, together and on our own terms, we can make our lives in this (still) attractive and vibrant neighbourhood better (*dass wir zusammen unser Leben in diesem eigentlich sehr reizvollen und [noch] lebendigen Stadtteil besser selber gestalten können*)" (Figure 4.4).[19] While negotiations began with municipal authorities, the occupants were eventually evicted in the early hours of the morning of 14 June 1977 in a large police operation that took place just hours before a scheduled ruling on the legality of the demolition (Böttcher et al., 1978: 67).

The eviction and destruction only confirmed the worst fears of local activists, many of whom later became part of the BI SO36 citizen's initiative. In an article in *InfoBUG* in the lead-up to the TUNIX Congress, former members of the Feuerwache Stadtteilzentrum reflected on the possibilities that the meeting might hold for radical housing activists working in Berlin and elsewhere in West Germany. The authors of the article spoke of the emotional letdown that accompanied the June eviction and reflected on the need to forge the kind of solidarities

Figure 4.4   The Fire Station on Reichenberger Straße in Kreuzberg, occupied in Spring 1977, cover of brochure produced in December 1977 (Papiertiger Archiv Berlin).

and spaces that TUNIX seemingly promised. "In the fire hall," they wrote, "we shared an intensive everyday relationship. This helped us to tackle our problems and was a testament to our political strength." "We still want everything as before," the text continued, "housing – politics – life. The neighbourhood social centre we dream of should encompass everything so that we don't die of boredom."[20]

The TUNIX Congress, if anything, only served to intensify these sentiments. It also acted as a new springboard for direct action and the escalation of the 'housing struggle' (*Häuserkampf*). The unwillingness of local municipal authorities in Kreuzberg to engage with BI SO36 compelled the group to adopt a more aggressive repertoire of contention. Whilst early accounts of the history of squatting in West Berlin (Laurisch, 1981) have focused almost exclusively on the role that housing scarcity played, it was clear that a turn to increasingly radical measures *connected* an urgent politics of housing to the range of tactics developed by the anti-authoritarian revolt of the 1960s and 1970s. In this context, BI SO36 operated a storefront in Kreuzberg on Sorauer Straße where residents could meet to discuss housing-related issues from rent hikes to eviction notices. They also

Figure 4.5 Bürgerinitiative SO36 Flyer announcing the occupation of two squatted apartment blocks in Kreuzberg, reproduced in the *Südost Express*, February 1979 (Papiertiger Archiv Berlin).

continued in 1978 to publish the *Südost Express* which traced and documented the relationship between housing inequality, urban renewal and property speculation in Kreuzberg and West Berlin more generally. Landlords were targeted by members of the initiative with letter-writing campaigns, including the city-owned BeWoGe (*Berliner Wohn-und Geschäftshaus*) housing association which had over 300 empty properties in Kreuzberg. Press conferences were held and local politicians were approached to little effect. Facing official intransigence, BI SO36 decided, by the end of 1978, to finally take the matter into their own hands. On 3 February 1979, two apartments in two separate buildings managed by BeWoGe were squatted by members of the initiative. As they made clear in a widely distributed flyer (Figure 4.5) that was re-published in the *Südost Express*:

> In our district, hundreds of apartments are empty and falling apart. Cheap apartments are demolished because landlords no longer put them up for rent. This is

against the law. On the 3rd and 4th of February, the citizen's initiative SO36 (after the local post code) wants to restore the lawful condition of rental accommodation. Starting at 10 o'clock we will occupy and restore one apartment in Luebbener Straße and another on Goerlitzter Straße.[21]

The squatters originally intended to use the occupations as symbolic sites of protest, drawing public attention to the effects of urban renewal and the acute shortage of housing in Kreuzberg. As the flyer suggested, under the "*Zweckentfremdungsverbots-Verordnung*" ("ZwVbVO") it was illegal for landlords to leave housing empty for more than three months (Laurisch, 1981: 37; Klein and Porn, 1981: 113).[22] At the same time, landlords could claim a 'legitimate interest' under the law and receive a license that guaranteed the property's vacant status. Financial sanctions were equally difficult to impose. In response to limited legal options, the squatters adopted the motto, "it is better to squat and mend than to own and destroy (*lieber instand(be)setzen als kaputt besitzen*)" (Laurisch, 1981: 34).[23] The occupiers also used, for the first time, the term *Instands(be)setzung* to describe their movement; the term itself a clever combination of the German for maintenance (*Instandsetzung*) and squatting (*Besetzung*). "We wanted to show," as Kuno Haberbusch a member of BI SO36 recollected, "that it is cheap to repair apartments. So we tidied up, painted, and repaired windows" (quoted in Rosenbladt, 1981: 36). For another squatter, "the term '*Instandbesetzungen*' [...] suggested [on the one hand] a concrete political engagement with property relations [...]. On the other hand, it also offered a new commitment to repair and maintenance." "This is," they concluded, "an expression of the fact that we, as a movement, offer a constructive alternative to redevelopment. The constructiveness of our approach thus lies in *what we do* with flats" (Klein and Porn, 1981: 112; emphasis added).

The nature of the occupations and their public popularity prompted activists to shift tactics as they increasingly eschewed the symbolic for the prefigurative. It was time, according to BI SO 36, to take matters into their own hands and intensify their actions even if this meant breaking the law.[24] New squats were quickly established. In March 1979, a group of youths occupied the fourth floor of an abandoned factory on Waldemarstraße.[25] In June 1979, the old UFA film studios in Tempelhof were squatted by a group of 80 activists who were eventually successful in legalising their occupation, turning the studios into a community cultural centre.[26] A few months later, Leuschnerdamm 9, another BeWoGe property, was squatted and in November of the same year, a further three buildings at Cuvrystraße 20, 23, 25 were taken over by members of BI SO36 (Laurisch, 1981).[27] In a press release published in the *Südost Express*, they reiterated their demands for "the immediate renting out of empty apartments" and the "renovation and modernization of properties according to the wishes of the tenants".[28]

The initial turn to squatting met, however, with little success. Whilst BeWoGe agreed to negotiate new leases for 40 apartments, the city's official renewal policy continued unabated. Empty flats were bricked up by city-owned and private

Figure 4.6   American soldiers conducting urban military exercises in Kreuzberg. *Südost Express*, December 1979 (Papiertiger Archiv Berlin).

housing associations whilst suspicious fires were started in a number of vacant properties. As a photo in the December 1979 issue of the *Südost Express* also made clear, the long-term abandonment of housing in Kreuzberg offered little more than a training ground for US soldiers stationed in Berlin to practise the latest urban combat manoeuvres (Figure 4.6). For the few residents still living in block 104 between Oranien-, Mariannen - and Skalitzer Straße, this meant waking up one morning in the autumn of 1979 to the sound of gunfire and a platoon of US soldiers armed with live ammunition smashing windows, breaking down doors and punching holes in the wall. "This is war," the residents concluded, "war against us."[29]

Activists in Kreuzberg responded to the escalating crisis in housing through a further wave of occupations as buildings at Heinrichplatz 14, Naunynstraße 77, 78, 79; Mariannenstraße 48 and Leuschnerdamm 37/39 were all squatted in the early months of 1980. On 28 March 1980, squatters representing nine different buildings met to form a squatters' council (*Besetzerrat*). "We seek to overcome," the new council announced in its first press release, "the isolation facing individual groups of squatters and foster greater cohesion and collaboration. Despite the diversity of our movement, we want to work together against an official policy of demolition and defend ourselves against attacks and evictions."[30] At the same time, though not known to activists until many months later, the Berlin police had begun to put together a special unit specially tasked with enforcing order and containing the rise of militant activism across the city.[31]

In the following months, increasing police brutality became a prominent feature in a number of operations as local authorities sought to criminalise the actions of squatters. An attempted occupation of a building on Wrangelstraße on 29 May 1980 resulted in a large police operation and the arrest of six squatters. In the aftermath of the eviction and under police protection, the building was gutted by a team of workers hired by the landlord.[32] A few days later, the first squat outside the SO36 postcode in Kreuzberg at Chamissoplatz 3 was cleared in another major police raid in which 16 people were arrested, prompting activists to ask whether it was such actions which ultimately represented the "new form of citizen participation?"[33] The squatters' council expressed serious concern that their activities were being targeted as part of a concerted campaign. "We are constantly under surveillance by plainclothes police," they remarked. "We are being photographed, stopped and searched, tracked, abducted and beaten." "The violence," added another squatter in a newspaper interview, "didn't come from us (*Die Gewalt geht nicht von uns aus*)."[34] Squatters also countered accusations that they were simply a front for extreme leftist organisations and drew attention to the alternative self-help infrastructure that they were building across Kreuzberg and which they saw as an urgent and necessary response to creative destruction and the urbanisation of capital. "All the problems," they concluded, "have been created by urban renewal and rampant speculation. They cannot be covered up with a police baton."[35]

While the number of squatted houses rose steadily to 18 over the course of the second half of 1980, it was the attempted occupation of an apartment building at Fränkelufer 48 on 12 December 1980 that served as the major turning point for the wider housing movement in West Berlin. Around 5 o'clock, a small group of housing activists gained entry into number 48 with a view to squatting the building. The apartment complex – facing the Landwehrkanal – was almost entirely abandoned and the squatters quickly moved onto the third floor. However, one of the few remaining tenants had notified the police who soon arrived on the scene and entered the house to arrest the squatters. The squatters offered no resistance and were taken away by the police to be processed at the Friesenstraße station a few blocks away. As the arrests were taking place, a large crowd – including many supporters and other members of Berlin's squatting scene – had gathered around the building and down the street towards the

Figure 4.7 Protests in the wake of the eviction of squatters from a house on Fränkelufer 48, Kreuzberg, 12 December 1980 (Michael Kipp/Umbruch Archiv).

Admiralbrücke. Rumours quickly circulated through the crowd that a squat down the road on Admiralstraße (no. 20), which had been occupied on October 24, was to be cleared as well. Barricades were thrown up in front of the building on Admiralstraße as the number of protesters swelled. The police had also brought in additional units and a decision was made to clear the protesters. Tear gas was fired into the crowd who were set upon by the police. The crowd dispersed towards 'Kotti' (Kottbusser Tor) and over the course of the next five to six hours a series of pitched battles between the police and the protesters were waged across Kreuzberg (Figure 4.7).[36] The night ended with 58 arrests (18 women and 40 men). Of those arrested, 19 were under the age of 21, 32 between the ages of 21 and 30, and seven were over the age of 30. Protesters were charged under Sections 123, 125 and 223 of the German Criminal Code (for trespass, resisting arrest, and assault respectively). One protester was run over by a police van and seriously injured and taken to hospital. There was damage to a number of storefronts across Kreuzberg (see discussion in Suttner, 2011; Ermittlungsausschuss im Mehringhof, 1981b).[37]

For members of the squatting scene, the violent actions of the police were seen as a deliberate provocation. The Berlin Senate had recently adopted a policy of 'pacification' which involved the tactical legalisation of certain squatted houses and the possibility for the 'swapping' of other 'illegally' occupied spaces with empty houses chosen by the Senate. In fact, apartments blocks at Admiralstraße 18b and 18d were part of a transfer scheme that had only been agreed upon on 11 December in conjunction with the Sozialpädogigisches Institut der Arbeiterwohlfahrt, a social welfare organisation (Suttner, 2011). As it turned out, it was officers stationed in front of these two blocks who were the first to arrive on the scene and arrest the group of squatters attempting to occupy Fränkelufer 48. The brutal crackdown on the night of the 12th also forced the hand of the Berlin Senate and they acted quickly to exonerate the actions of the police. The squatting movement was accused by the Senator for Housing, Harry Ristock, of trying to disrupt and sabotage negotiations. It was even suggested in a press release issued by the Kreuzberg municipal authorities that the squatters were never interested in a peaceful solution to the housing crisis.[38] Newspapers on the following day reported that the police had in fact prevented an occupation at Admiralstraße 18b and 18d (the empty houses designated in the transfer scheme), not Fränkelufer 48. The only evidentiary source provided was the police's own press release. This pernicious choreography – from accusation to misrepresentation – was further refined in the days to come as subsequent newspaper articles claimed that the attempted occupation on Admiralstraße had been followed by the events on Fränkelufer.

Squatters responded through further escalation. On 13 December, an impromptu march was held on the Kurfürstendamm ('Ku-Damm') which led to another confrontation with the police. Numerous arrests were made and a number of stores were vandalised over the course of the evening.[39] The following day, Fränkelufer 48 was again squatted. This time the police made no effort to prevent the occupation. The squatting movement also insisted that any further negotiations with the Berlin Senate and local municipal authorities in Kreuzberg were contingent on the release of those already in prison. On 15 December, further demonstrations were held in solidarity. Over 3000 protesters marched down the 'Ku-Damm'. Street battles ensued and a number of banks and supermarkets were once again damaged. A peaceful demonstration – the largest yet and numbering 15,000 – was held on 20 December in front of the main prison in the district of Moabit.[40]

Little progress was made, however. On the one hand, a small circle within the wider squatting scene had begun to question what they saw as an increasingly dogmatic approach within the movement that favoured antagonism and separation over constructive engagement. Appeals were made for a negotiated settlement with the Senate.[41] On the other hand, the squatter's council made it clear in a statement released on 16 January 1981 that a housing crisis that had condemned significant numbers of people to misery required new autonomous forms of housing and shelter. To occupy, in their view, was to seek out spaces for self-determination and emancipation. "We want to determine our lives at home just as we do at work (*wir wollen unsere Lebensbedingungen am Wohn- wie am Arbeitsplatz selbstbestimmten*). We believe that squats are a disruptive possibility, an imaginative cog in the wheel of the state."[42] At stake here, so it seemed, was the articulation of a rather different understanding of the city. If squatters protested against decay and scarcity, they also foregrounded the importance of *re-appropriating space* for the production of a "transformed and renewed right to urban life" (Lefebvre, 1996 [1967]: 158; see Vasudevan, 2011a, 2014a).

It is perhaps not surprising, therefore, that the events of 12 December and its aftermath have occupied such a prominent place within the history of the Berliner squatter scene and the wider autonomous movement in West Germany. As one report concluded, it was the night in which order was "turned head over heels in Kreuzberg".[43] For many squatters, the riots constituted a transgressive event in which existing social boundaries collapsed and new connections and possibilities were forged (Häberlen and Smith, 2014: 634). The period that immediately followed thus became a high point for squatting in Kreuzberg and West Berlin more generally as the scene grew rapidly and moved into other neighbourhoods with similarly old and empty housing, including Schöneberg, Charlottenburg, Moabit, Neukölln and Wedding (Laurisch, 1981; Suttner, 2011; Vasudevan, 2011a). By 6 February 1981, over 50 separate apartments blocks were occupied across the western half of the city. Three weeks later the number had risen to 100. Occupations peaked on 15 May 1981 at which point 169 houses were occupied in West Berlin. From the beginning of 1979 until 27 January 1982, over 239 houses had been squatted (some three or four times). In that same period, the police were involved in 196 separate 'incidents' involving occupied houses and, according to their records, 2289 people were identified as squatters (see Suttner, 2011).[44] There was considerable public sympathy for the new squatting movement. In 1981, the prominent polling agency, Allenbach, conducted a survey which showed that over 53 percent of the West German public believed that squatters were right in their criticism of urban development and "clear-cut renovation (*Kahlschlagsanierung*)" (Reichardt, 2014: 500; Korczak, 1981: 27).

Growing public criticism of the city's urban regeneration programme also coincided with a major finance scandal that would eventually lead to the resignation of the mayor, Dietrich Stobbe, in January 1981. The so-called 'Garski affair' drew further attention to the corrupt connections between the West Berlin Senate and the construction industry. In 1978, the city opened a line of credit for architect and housing contractor Dietrich Garski at the local Bank of Berlin to support a building project in Saudi Arabia despite a lack of supporting documentation. The project fell through and, as payments fell into arrears, Garski went into hiding. The city and the West Berlin taxpayer were left to foot a bill of over 110 million DM (see Laurisch, 1981). Whilst squatters undoubtedly took advantage of the political instability caused by the Garski scandal to occupy empty buildings across Kreuzberg, Schöneberg and other neighbourhoods in the city, they ultimately did so with a view to creating a radical infrastructure that encompassed spaces which challenged dominant understandings of home, family and work. In other words, to claim the right to a different city was not limited to the seizure and occupation of buildings. It centred on what those buildings might become as spaces that promised a different form of life than the one that the occupiers "[were] compelled to live" (Brandes and Schön, 1981: 69).

Making 'Free Space': The Material and Emotional Geographies of Squatting

"But then, one morning at half past six, we simply went inside," wrote the *taz* reporter Benny Härlin in an article in the journal *Kursbuch* as he recounted his first moments as a squatter in West Berlin in the early months of 1981 (1981: 5). Härlin's detailed account of his life as a squatter offers important insights into the everyday practices of occupation and the repertoire of skills and tactics adopted by activists in West Berlin. For Härlin and many other squatters, the very act of occupation constituted, in anthropological terms, a *liminal* experience as they crossed a threshold between their past and future lives (see Turner, 1969). "Going in," according to Härlin, was much easier, in the end, than the planning process that preceded it. "The composition of our group," added Härlin, "fluctuated as we met on three separate occasions in order to plan and get to know each other. Some backed out, others brought friends with them. Each of us probably thought at least once that the whole thing was just a nice dream. Who would really move in or who would rather retreat to the status of 'supporter' was at the time of the occupation unclear, as was our idea of what should happen to the house" (1981: 5). Nevertheless, the squatters were well-prepared, bringing with them pipes, tools, torches and replacement locks. The first thing they did was to change the locks and drop banners from the windows publicising the occupation in order to secure the space and prevent a swift eviction by the police. The squatters quickly moved through the empty rooms, choosing the best-preserved apartment which they began to clean up in order to hold a press conference. Others went out onto the street distributing leaflets to passers-by (1981: 5).

Anticipating more recent forms of urban exploration (see Garrett, 2013), the early moments of an occupation were described by Härlin and many others in pioneering expeditional terms. Härlin recalls "a triumphant but also oppressive feeling, to wander for the first time through the long, echoing hallways." "You are an intruder," he continued, "a stranger in this dead world" (1981: 6). If the house appeared, on first inspection, to be 'dead', it was slowly brought back to life, as Härlin argues, through a precarious process of accretion and assembly as materials and infrastructures were incrementally added to satisfy new needs and possibilities (Vasudevan, 2014b; see McFarlane, 2011): "At first, we installed ovens in just one apartment. That was the scouts' era, with little more than candles and water in canisters from the neighbours, ten people in our sleeping bags crammed in a single room." The process of restoration was, in turn, painfully slow. In Härlin's own words,

> Every new achievement was celebrated. The zenith of luxury and comfort: a bathroom was set up after six weeks. A bit of everyday life is finally moving in (*Ein Hauch von Alltag zieht ein*). First comfortable corners are forming close to the ovens. A little shelf, a candlestick, the first bed linens instead of the sleeping bag are indicating the extension of privacy. The number of inhabitable rooms is now sufficient for two persons to retreat for the night (1981: 7).

While the process of mending and repair documented by Härlin spoke to an incremental form of urban dwelling that responded to basic needs, squatters also saw the same buildings in light of the creative possibilities they offered. In their own eyes, the built form was never simply the container or the context for the creation of new experimental geographies. If anything many, if not most, squatters took as axiomatic the active materiality of a building as a necessary condition for experimenting with "new forms of collective living" (Kunst & KulturCentrum Kreuzberg, 1984: 13).

For Härlin and others, the very act of occupation was therefore understood as a form of 'resuscitation'. In practical terms, this depended on a modest ontology of mending and repair as squatters often confronted abandoned spaces which required significant renovation (McFarlane and Vasudevan, 2013; see also Vasudevan, 2011a). Makeshift materials and do-it-yourself practices combined with the sharing of food and other resources to provide the material support for collective self-management. As Härlin noted, "our rule was 'learning by doing'" (1981: 8; English in original). Occupation and renovation thus moved in a series of stages. As squatters entered into a vacant building, the first task was to remove any immediate dangers. Fallen debris was cleared while dry rot and mould were identified and removed from apartments. Once this was done, squatters secured rooms that could be immediately occupied, clearing further debris and garbage from apartments and courtyards. Only then did they turn to the task of repairing structural problems that were a result of the owner's neglect and, in many cases, deliberate vandalism (see MacDougall, 2011a: 148). Härlin recalled how old manual handbooks such as *Der Heizungstechniker* (*The Heating Technician*, 1929) and *Die Neue Bauwirtschaft* (*The New Construction Industry Manual*, 1946) did the rounds between occupied houses as windows were repaired or replaced, old and exposed electrical wiring was fixed, and proper plumbing was restored (1981: 8). After this, squatters turned their attention to damaged floors and roofs. Repairs would often take months and, in some cases, years to complete.

The squatters who thus occupied three apartments on Cuvrystraße on 26 December 1979 encountered spaces "full of garbage and refuse" (Laurisch, 1981: 68). One apartment had been completely walled up by the landlord. With help from other members of BI SO36, the squatters were able to clear one room on the first evening. Over the next few months, the houses were painstakingly renovated. Electricity meters were acquired from the local utility company, new wastewater pipes were installed and existing water pipes were repaired. In the early summer of 1981 a new roof was finally completed (Laurisch, 1981: 103, 106; see also Kerngehäuse Cuvrystraße, 1980). Repair and rehabilitation also depended on the sharing of materials, practices and know-how between squatters and their supporters across Kreuzberg and West Berlin (Härlin, 1981). Occupying spaces and setting up new housing projects required, after all, a series of specific skills and training. As the squatters who occupied a building on Manteufelstraße pointed out, once the "junk was out" and they were in, a long and difficult road of labour intensive work began.[45]

It was, in fact, the squat on Manteuffelstraße which set up the Bauhof Handicraft Collective (*Bauhof Handwerkskollektiv*) as a place where squatters and other locals could learn basic construction skills and techniques from other squatters, many of whom had completed an apprenticeship in a trade or were attending a vocational school in West Berlin.[46] The *Bauhof* took on a key coordinating role within a wider activist milieu in Kreuzberg that encompassed a growing number of handicraft collectives which had emerged during the 1970s as part of an expanding network of radical self-help activities (see Reichardt, 2014). The *Bauhof* supplied squats with inexpensive building materials that were either recycled or purchased cheaply in bulk. It also ran a series of articles in the *Instand-Besetzer-Post*, one of the chief publications of the squatting scene, that provided detailed DIY instructions on a host of repair-related issues.[47] Ultimately, the *Bauhof* became a site where squatters and architectural professionals often converged to discuss and experiment with new and innovative approaches to participatory design and adaptive re-use (Figures 4.8 and 4.9).

In this way, squatters in West Berlin in the early 1980s were able to cultivate an ethos of self-determination and self help – a radical DIY empiricism – that transformed the ways in which people, materials, ideas and resources came together (Simone, 2010: 5; see Vasudevan, 2014b). Squatters were also acutely aware, in this context, of the complex set of actions, connections and forms that sustained the spaces they lived in. "Well, on such premises, there is always something that is not working," noted one squatter in 1981. "And if there is nobody else," they added, "you develop a kind of relation to it which you could call responsibility. You know where the cables run, as much as you know the sounds of the environment, the people, *it is all interconnected*" (Härlin, 1981: 18; emphasis added). A recognition of the often improvised makeshift nature of occupation did not limit, however, the ambitions of squatters. For many, if not most, squatting offered a "potentially different lens for linking everyday life, uncertainty and the possibilities of [an] alternative urbanism" (McFarlane, 2011b: 163). At stake here, as Ash Amin (2014) has argued in a related context, is a particular relationship between collective organisation and radical infrastructure and between a politics of bricks and pipes and a politics of contestation and resistance (20).

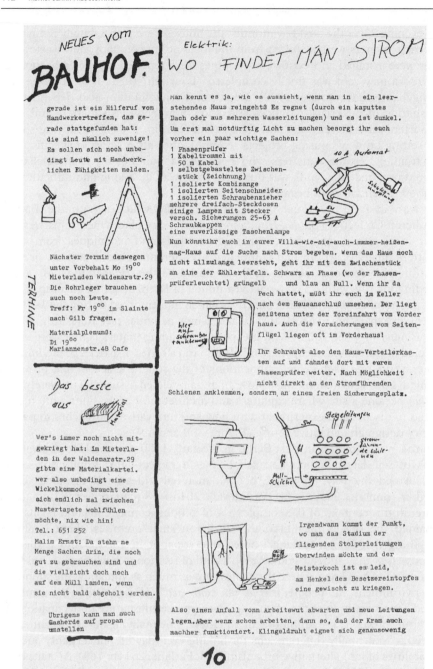

Figure 4.8 DIY instructions in the *Instand-Besetzer-Post,* 17.3.1981, p. 10 (Papiertiger Archiv Berlin).

Figure 4.9   The makeshift city: Rehabilitating occupied spaces in Kreuzberg (Umbruch Archiv).

The *material geographies* of squatting were inextricably tied, therefore, to a broader struggle to re-imagine city life as a shared political project. Whilst the practices that were developed by Berlin squatters undoubtedly spoke to pressing housing needs, they also drew upon a protest repertoire that was inspired by the actions of the anti-authoritarian left of the 1970s (see Brown, 2013; Reichardt, 2014). To live their lives "in ways otherwise not possible", squatters actively sought to create spaces that *prefigured* how one might come to know and live the city differently (quoted in MacDougall, 2011a: 150). Squatters thus responded to normative assumptions about living and the 'home' through the questioning of its more basic spatialities. "We didn't just occupy buildings," they argued, "we occupied the substance of buildings (*Wir haben nicht nur Gebäude besetzt.Wir haben Inhalte von Gebäuden besetzt*)" (Freiburg squatters quoted in Haller, 1981: 11). This was a process that took on a number of forms and embodied a broad range of spatial practices which differed from one squat to another. The degree to which the Berlin squatter scene (*Besetzerszene*) cohered as a single urban movement was, in other words, open to debate. For many activists, the occupation and re-appropriation of empty buildings and houses offered a direct challenge to urban speculation, widespread housing shortages and commercial planning initiatives. As a form of 'direct action', squatting represented, according to this view, both an "attack on the unjust distribution of urban goods" and an attempt to link alternative forms of collective living with non-institutional grassroots urban politics (López, 2013: 871; see Amantine, 2012; Dellwo and Baer; 2012). For others, squatted houses served as emancipatory sites that would come to challenge traditional identities and intimacies (Bodenschatz et al., 1983; Péchu, 2010; SqEK, 2013; Vasudevan, 2011a). This was often predicated on *queering* the home as a site of domesticity and social reproduction and where the everyday micro-politics of making a 'home' countered not only traditional performances of housekeeping and kinship but also unsettled conventional distinctions between publicity and privacy and, in so doing, proffered radically new orientations for shared living (Brown, 2007; Cook, 2013; Amantine, 2011). And for others still, the act of 'occupation' simply served as a pretext for producing a suitable arena for a wide range of subcultural practices. In the case of many punks, for example, little effort was made to renovate squatted spaces. One would simply move on when the housing conditions became unbearable (see Bock et al., 1989).

At the most rudimentary level, architecture still served in most cases as a guiding frame for the creation of new sustainable structures of organising, working and living (Feigenbaum et al., 2013: 41). If the traditional Berlin tenement house (*Mietshaus*) was widely seen as a key symptom of creative destruction, for squatters it also offered an alternative to the "organized inhumanity" of the modernist tower block.[48] As the historian Brian Ladd once noted (1997: 102), "a hallmark of the *Mietskaserne* [is] its flexibility". The existing building structure only encouraged squatters to maximise communal spaces. Over half of all squats in West Berlin contained collective self-managed spaces as the permeability of buildings was increased and re-engineered to suit the changing needs and wishes of their inhabitants. Therefore, walls were removed in order to increase the size of social spaces including kitchens, while stairwells were created to produce a new geography of movement through buildings, now connected and held together by an interspatial network of doors, passageways, courtyards, and vestibules (Sheridan, 2007: 115). Changes to the inner layout of squatted spaces were also accompanied by the transformation of their outward appearance. Hand-painted banners were draped from windows while the walls were colourfully painted acting, in the words of some squatters, as "urban tattoos" that provided, in their eyes, "a stark contrast to the emotionally vacuous visual landscapes of normal urban life" (Anon, 1981: 41–2; Häberlen and Smith, 2014: 631). Echoing more recent calls for a "critical geography of architecture" (Lees, 2001; see also Adey and Kraftl, 2008; Datta, 2008; Jacobs and Merriman, 2011; Vasudevan, 2011a), these experiments with the built form became a key process for exploring a new micropolitics of alignment, interdependency, and connection (Simone, 2004: 12).

In the end, each squat adopted its own approach to the organisation of everyday life and the development of shared spaces. Separate floors were, in some cases, designated for different activities (with workspaces on one floor and bedrooms on another). In many other squatted spaces, walls were torn down in order to create communal kitchens and dining rooms. In other houses, large common rooms were designed and used for general meetings (*Plena*) and other cultural activities including parties, gigs, theatre performances and films. The micropolitics of squatting drew, in this way, on "modes of spatialisation" that placed particular emphasis on a process of experimentation and re-appropriation (Guattari and Rolnik, 2008: 64). At the same time, they were also connected to wider geographies of care and solidarity that were embedded within the local neighbourhood. Across West Berlin, and in Kreuzberg in particular, squatters responded to a lack of social services by developing new social spaces that were "intended to address the needs of the existing residents, provide an improved infrastructure, and foster better neighbourly encounters" (MacDougall, 2011a: 150). "Our house concept," according to one of the occupants of a squat at Manteuffelstraße 40/41, "is designed to help shape and maintain the living environment to best serve all neighbourhood residents."[49] Taken together, the broad spectrum of social spaces developed by squatters in West Berlin played a significant role in the consolidation of an alternative infrastructure that was constitutive of both localised spatialities and wider articulations of autonomy and self-determination (see Featherstone, 2012). Such "reciprocities of infrastructure and sociality" assumed a number of forms including alternative cafés (Café Krantscho, Lokal Lummerland and Das Besetzereck), concert venues (Bobby Sands Pub and Café Knüppel), daycares, workshops and youth centres as well as a children's petting zoo, a Turkish bath, and a homeopathic clinic (Amin, 2014: 13; Reichardt, 2014: 546–547).[50] Some of the larger squatted spaces, including the Regenbogenfabrik on Lausitzerstraße and the Kerngehäuse on Cuvrystraße, became social centres in their own right. Both sites recalled an earlier history of mixed-use in Kreuzberg while hosting a wide range of initiatives that blurred the boundaries between "new forms of living" and "new ways of working" (see Kerngehäuse Cuvrystraße, 1980: 4). For the squatters who occupied an abandoned chemical factory now known as the Regenbogenfabrik, 'occupation' represented far more than an act of rehabilitation and renovation. It was also widely understood as an ongoing 'process' through which a meaningful social infrastructure was assembled, sustained and extended. As was the case with many squats, the Regenbogenfabrik placed particular emphasis on horizontal decision-making and collective self-management and applied these practices to both the day-to-day life of the squat and to projects that supported a larger community (cycle repair shop, carpentry workshop, daycare, cinema, hostel and community 'info shop') (see Regenbogenfabrik, 2006: 10, 19) (Figure 4.10).

In other instances, more trenchant forms of occupation were mobilised with a view to creating particular spheres of identification that would encourage forms of interaction "that other deliberative spheres [would otherwise] constrain or censor" (Jackson, 2000: 17). For example, the former chocolate factory of the firm Greiser und Dobritz at Mariannenstraße 6 and Naunystraße 72 in Kreuzberg was squatted in 1981 by a group of women who "were looking for rooms where they could live undisturbed and meet freely with each other without the unwanted attention of men and without being restricted solely to their own private apartments" or the routinising

Figure 4.10   The Regenbogenfabrik social centre in Kreuzberg (photo by author).

spatial demands of domesticity and social reproduction (Bosse and Zimmer, 1988: 10).[51] Renovation and modernisation of the Schokofabrik was undertaken by the occupants and focused on a process of participatory architecture and sustainable redevelopment. The Schokofabrik was one of many spaces occupied by women in response to the patriarchal attitudes that continued to subsist within the radical left in West Germany. For many women, squatting only reinforced the view that "structural patriarchal violence has many multifaceted faces both subtle and crude" (quoted in Amantine, 2012: 36). In the words of one commentator writing in the *Instand-Besetzer-Post*, "well, now we live in squats. But paradise is still so far away. For behold, the men are still the same. We are not only suppressed by the State. When we get home, we are therapists and 'mummies' not squatters."[52]

The emergence of an autonomous feminist movement in the 1970s provided inspiration, in this context, for a number of women-only initiatives. On 5 January 1981, a group of 12 women connected to the autonomous and lesbian scene in West Berlin occupied a house on Liegnitzerstraße which soon became known as the *Hexenhaus* ('witches' house'). "It is well-known," the squatters proclaimed, "how difficult it is to find an apartment in Berlin. For us women, it is almost impossible whether we are alone or a group. We have, therefore, occupied a house with and for women."[53] The Feminist Women's Health Centre (*Feministische Frauengesundheits-Zentrum*) which was founded in Berlin in 1974 also moved into the building. Over the next few weeks, houses on Mariannenstraße and Naunynstraße were squatted by women who "despite differences [...] knew that we wanted to live together as a group" (see Amantine, 2011: 90–91).[54] In February 1981, a women's-only café was set up in a mixed squat in Berlin-Moabit and it soon served as a key meeting point for a network of autonomous feminist spaces across West Berlin (Figure 4.11).[55] At the same time, The Women's Research, Education and Information Centre (*Frauen Forschungs-, Bildungs - und Informations-Zentrum*) was created in a squat in Charlottenburg at Dancklemanstraße 15. Finally, a house in Kreuzberg on Kottbusser Straße was briefly occupied by eight women in response to the lack of affordable housing for Turkish women and their children (see Amantine, 2011).[56]

Squatting thus played an important role in the articulation of new identities and the further development of the feminist movement in West Berlin and elsewhere in West Germany. It also became a site from which a new sexual politics and a wider queer sensibility emerged and took hold. The occupation of Potsdamer Straße 139 in April 1981 by local sex workers and other activists should be seen in this context, and was part of a broader project that included the establishment of living spaces for sex workers and educational and social assistance for their children. A few months earlier, the first 'Queen House' (*Tuntenhaus*) was set up in an abandoned wing of an apartment at Bülowstraße 55 in Berlin-Schöneberg where the occupants established a little community to share their "warm love and work".[57] As the writer and housing activist, Amantine, argues, "the squatting of houses by gay and lesbian activists, queers and trans groups – their visibility, perceptions and politics – led a movement inside houses to increasingly question firmly entrenched gender norms and [...] the forced imposition of heterosexual ideas, structures, and kinship networks" (2012: 38).

In general, the emergence of the squatting scene in West Berlin was characterised by its sheer diversity, attracting students, local youth, apprentices, runaways, drop-outs, hippies, spontis, anarchists, punks, gay and lesbian activists, radical queer and trans groups, migrants, refugees, and environmentalists (see Katsiaficas, 2006). There remains, however, little reliable data on the social composition of the squatting scene (Holm and Kuhn, 2011: 647). According to an article published in *Die Zeit* in August 1983, 65 percent of squatters were men, 90 percent were German, 78 percent were single, 35 percent were under the age of 21, 40 percent were between the ages of 21 and 25, 36 percent were schoolchildren or students, 26 percent were in employment, and 38 percent were without a recognised job.[58] An official police report showed that of the 692 squatters arrested in 1981, over 82 percent were between the ages of 18 and 30.

The future is fem@le

FRAUENCAFE
TERMINKALENDER DER JAGOWStr.12

Freitag 12.6. 20°°   Mittwoch 8.7. 20°°
Zini brennt   Das 2.Gewachen der
                           Christa Klages
Mittwoch 17.6.20°°   Mittwoch 22.7. 20°°
Deutschland   Schade das Beton
bleiche Mutter   nicht brennt
Mittwoch 24.6.20°°   Samstag 23.7. 21°°
Alles hat hier   Frauendisco
seinen Preis

GLEICHBERECHTIGUNG
Gesucht
Liebevolles
Emsiges
Inniges
Chices
Herzchen
Betont
Ergeben
Rechtschaffen
Ehrlich
Charmant
Hilfreich
Tüchtig
Interressiert
Ganztags
Unterbezahlt
Nett
Gegängelt zu den

Betrifft NOCHMAL
BESETZERINNEN

Um endlich die alles zersetzenden Ge-
rüchtebrodeleien ein für allemal zu be-
enden: in obigem Garten findet nun endlich
das schon langangekündigte Besetzer'innen
treffen statt. Da is nix mit verschoben oder
ausfallen! Auch nicht wegen - wie man mir
z.B. weismachen wollte - irgendsoeiner Demo
in Heidelberg oder sonstwo. Die Lima-män-
ner sind bereits in Urlaub
geschickt. Deshalb gilt jetzt:

FRAUEN KOMMT
am 12.13.14.6. in
die LIMASTR.29
Zehlendorf. Nahe Mexikoplatz
Bringt Schlafsäcke Instru-
mente + Essachen mit.

Ein Beitrag aus der Unabhängigenfah.
FRAUENSEITE?

Na, ja - nun haben die Frauen auch in der BP ihre
Seite, ihre Ecke, wo sie sich austoben können -
na schön - werden wa eben schreiben aus der Haus-
bäsetzerinnenbewegung, die ja nicht enôstehen darf,
von wegen der 'Spaltungsgefahr'.
Warum überhaupt eine Frauenseite? Sind die Saus-
besetzer auf mal so frauenfreundlich geworden?
oder steckt was anderes dahinter? - hat wahrschein-
lich wieder Alibifunktion - kennen wa schon!
Wir haben keinen Bock und wieder mal in dieses
Ghetto abschieben zu lassen und diese angeblichen
Frauenprobleme, die ja eigentlich die Macker haben
(es nur nicht wissen und nicht merken d.Sätz)
wieder mal als unsere Probleme aötun! Wir setzen
voraus, daß gerade linke Männer nicht nur für ihre
alternative Lebensvorstellung arbeiten und kämpfen,
sondern sich auch unbequeme Gedanken machen müssen
bezüglich ihrem Mackerverhalten. Wir sind nicht mehr
für Euren unreflektierten Chauvinismus verantwort-
lich, Hausbesetzerin zu sein reicht und, demxxxgux
Krankenschwester und Therapeutin könnt ihr woanders
suchen. deshalb fordern wir: keine mickerige Frauenseite, sondern eine r a d i k a l   f e m i n i s t i
s c h e   Besetzerpost !!!

A.S.H.

Die einzigen Frauen, die jemals Beiträge für
die Frauenseite geliefert haben sind also sowie
gerade gegen die Seite. Davon ist diese Seite
wie auf die Lehre von der Seite und wer etwas
dagegen hat soll dies mit Wort + Tat begründen.
23

Figure 4.11   Intimacy and utopia: The Frauencafé in Moabit, *Instand-Besetzer-Post*, 12.6.81, p. 23 (Papiertiger Archiv Berlin).

Over 38 percent were students.[59] While political goals and views undeniably differed, the squat was still seen by most as a place of collective world-making: a place to imagine alternative worlds, to express anger and solidarity, to explore new identities and different intimacies, to experience and share new feelings, and to defy authority and live autonomously. Squatting offered an opportunity to quite literally build an alternative habitus where the very practice of 'occupation' became the basis for producing a common spatial field, a field where principles and practices of cooperative living intersected with juggled political commitments, emotional attachments, and the mundane material geographies of domesticity, self-governance and renovation. At stake here, in other words, were practices that transformed the act of squatting into what Sara Ahmed has elsewhere described as a phenomenological 'orientation', an orientation that involves different ways of extending bodies, objects and practices into space while producing a world of "shared inhabitance" (Ahmed, 2007: 3).

If the history of squatting in West Berlin can, in this way, be understood as a series of spatial performances and practices aimed at creating liveable places that are themselves tied to particular forms of activism, empowerment, and self-determination, these were struggles that were conditioned and shaped by a complex constellation of affects, emotions, and feelings. This is hardly surprising, perhaps, given the radical left's growing commitment in West Germany to challenging what they saw as "the emotionally crippling effects of capitalism" and the feelings of anxiety, boredom and loneliness that , in their eyes, it produced (Häberlen and Smith, 2014: 617). As recent theorists of social movements as well as a number of geographers have moreover argued, "emotion is fundamental to political life and always a factor in the realm of activism, something that stirs, inhibits, intensifies, modulates, impedes, incites" (Gould, 2009: 439; see Brown and Pickerill, 2009; Brown and Yaffe, 2013; Feigenbaum et al., 2013). Activism, according to the sociologist Deborah Gould, traded, unsurprisingly perhaps, in feelings like anger, rage, indignation, hope, joy and solidarity but also those that might be less perceptible such as fear, shame, embarrassment, guilt and despair (2009: 2). At the same time, sites of activism were intensely *affective spaces* that also generated experiences which were bodily, sensory, inarticulate and nonconscious (Gould, 2009: 20; see Berlant, 2011; Massumi, 2002). "A focus on affect," adds Gould, "reminds us to take the visceral and bodily components of politics seriously, preserving for the concept of 'experience' some of its *felt* quality, some of the sense of being moved by the world around us" (2009: 441; original emphasis). To create a sustainable culture of resistance rests, therefore, on the "realization that one can ultimately never separate questions of the effectiveness of political organising from concerns about its affectiveness" (2009: 142, 143). For Stevphen Shukaitis, it is the very intensity of these connections and relations that are central to the politically radical character of care, resistance and solidarity. "We all too often fail," he concludes, "to appreciate the on-going work of social reproduction and maintaining community that these acts entail" (2007: 144, 142).

Squatters in West Berlin were acutely aware of the emotional and affective resonances of 'occupation'. The German term for occupation, '*besetzen*', also carries a psychoanalytical meaning, describing the process of investing mental or emotional energy in a person, object or idea. The act of squatting a house was linked, in this sense, to the psychological work that went into establishing an emotional space which offered a return to forms of community and neighbourliness that had been putatively erased by modernist planning initiatives and large-scale urban development. In other words, squatters did not just produce spaces. They also generated *emotional geographies* that helped to build and sustain an alternative urbanism, often in the face of considerable opposition. These geographies took on a number of different forms as squatters occupied spaces in which the "ongoing interactions of participants" produced sentiments, ideas, values and practices that encouraged and nurtured alternative modes of being (Gould, 2009: 178). As the Schöneberg *Besetzerrat* proclaimed in September,

> We did not squat to simply secure housing. We wanted to *live and work together* again. We wanted to put an end to the separation and destruction of communal living. Who in this city is not familiar with the *agonising loneliness* and emptiness of everyday life that has arisen with the ceaseless destruction of traditional relationships wrought by urban renovation and other forms of urban destruction? …. Many of us have, *for the first time, found a true home (Heimat)* in the squatted houses …. In the houses, we are trying to find something that does not exist anymore in society: *Relationships and Hope* (Ermittlungsausschuss im Mehringhof, 1981b: 6).[60]

The emotional work of squatting was undeniably intense. Indeed, the many descriptions of what it felt like to participate in the occupation of an empty building recall Emile Durkheim's (1995) notion of "collective effervescence". Durkheim used the term to convey "the transports of enthusiasm" and the "sort of electricity" that were shared, in his view, between a group of people who came together in common cause (Durkheim, 1995: 217; see Brennan, 2004; Gould, 2009). As the squatting scene exploded in the early months of 1981, each new squat in West Berlin was greeted with exhilaration and elation that stemmed in large part from a sense of purposefulness and political efficacy. What this entailed, as the countless posters, flyers and magazines produced by squatters during the period made clear, was a *displacement* of political action onto the lived rhythms and material foundations of daily life from work and leisure to housing and family. Autonomy and self-determination were actively assembled as new principles of association and cooperation were extended deep into the structures of everyday life (see Ross, 1988: 5, 33). All of this led, in turn, to the development of alternative forms of political encounter and gathering characterised by both a spatial openness and a spontaneous and immersive sense of time (Figure 4.12). The constant assemblies, meetings, performances, gigs, open houses, demonstrations, eviction resistances, etc. allowed squatters to move out of a society shaped by the logics of competition and consumerism and into an alternative city that still held, in their eyes, the potential for social transformation.

Former squatters reflected positively on the ways in which they were, on the one hand, able to forge an emotional field of commitment and solidarity and produce what Richard Sennett in a related context has described as a "new kind

Figure 4.12   "The city as festival": Street performance in Charlottenburg on 21 September 1981 (Wolfgang Sünderhauf/Umbruch Archiv).

of warmth" (1970: 7). On the other hand, they also drew attention to the negative consequences – the 'grind' – of shared living. In the words of one former squatter: "life [...] was often difficult, external and internal 'enemies' had to be confronted. There were tears and some comrades and principles had to be left behind. The motto, 'to live and work together,' led to a delicate balancing act between happiness and emotional breakdown which at times was rather sobering" (Regenbogenfabrik, 2006: 26). Another former squatter, who moved to West Berlin in the early 1980s to study, talked about the strong sense of loss that accompanied her choice to squat and abandon existing friendships and familial dependencies.[61] Others described how the intense affective atmosphere of a squatted house was often less a product of political activism than the everyday negotiation of shifting subjectivities. For many, everyday life *inside* a house was suffused with 'outside' politics as sectarian political divisions were quickly mapped onto the performance of daily activities. Indeed, for some, it was difficult to even imagine "sharing a bathroom and a kitchen with someone who didn't think the same way...[as one] did" (quoted in Davis, 2008: 269).

Squatted spaces were, in this way, both sites of liberation and possibility and sources of intense conflict and struggle. For many, these were spaces of cooperation and collective action where the "dream of self-determination" and the "symbiosis of living and working" were fulfilled (Regenbogenfabrik 2006: 5). And yet, the act of occupying was always a precarious process punctuated by continuous deliberation, disagreement and dissent. Disputes often centred on the organisation of squats as 'homeplaces' that sought to balance alternative forms of living with ongoing acts of protest and resistance (Juris, 2008; see Feigenbaum et al., 2013). In many squats in West Berlin, the pleasures and intensities of home-making were accompanied by conflicts and tensions over the organisation of everyday life. The division of labour – who was responsible for particular tasks and when (cleaning, cooking, mending, etc.) – was a source of heated negotiation and was compounded by the fact that many squatters were forced to combine their activism and their commitment to renovation work with full-time jobs. Such tensions also highlighted the gendering of roles within squats which drew considerable criticism from many women who felt that they were still largely responsible for all the major household activities (see Amantine, 2011). The squat, they lamented, was not only a space of radical alterity, but one where familiar hierarchical disputes were reproduced. As one squatter noted in anger, it simply "wasn't enough to whip up a pot of spaghetti once a week with the noodles so overcooked that they got stuck in your throat, or to play with kids only when they were not crying and had clean diapers".[62]

It is not surprising that the everyday life of squatting took its toll on the Berlin squatter scene. In many cases, the 'warmth' and good feelings simply did not last as new micropolitical arrangements gave way to personal animosities and conflicts. According to one squatter writing in the radical magazine *Der schwarze Kanal*, "in the very beginning, there was curiosity and the prickling excitement of new adventures. Everything was, if possible, to be done at once, that's how much fun we had with everything we did. Getting to know each other was just as much fun." But then, things changed as everyone fell back on old habits from "a complete indifference to what was happening around the house to a kind of laddish chauvinism that made you want to gag". As the author recalled, no one was willing to listen to each other and problems and tensions were simply papered over. "To be honest," they (somewhat ironically) concluded, "I can no longer be asked to live alone."[63] There is, of course, no predetermined connection between intense activism and activist burn-out, depletion and exhaustion. After all, the emergence of new geographies of solidarity and experimentation in the wake of the TUNIX meeting and the concomitant occupation of countless apartment blocks in West Berlin was widely seen by the radical Left in West Germany as a source of rejuvenation. At the same time, the tensions within houses (endless meetings and deliberations over household chores and finances, protest tactics, etc.) were compounded and amplified by the pressures exerted from both within (petty crime and drug use) and without (criminalisation) as squatters became increasingly divided over wider movement goals.[64] "There were many reasons other than the amount of work that made our life difficult," concluded one occupant of a squat on Kopischstraße. "Among our group of twenty," they added, "[…] it was almost impossible to find a consensus. There was a clash of different views especially as we discussed the future of the building."[65] Such concerns were shared by many other squatters. Should they negotiate with local authorities to secure legal status and funding to support renovation and rehabilitation? What role should violent resistance play in the face of heightened criminalisation and repression?[66] As one commentator in the *Instand-Besetzer-Post* concluded,

> Our situation in the squatted houses is therefore difficult. On the one side, there is the pressure from the outside: the criminalisation, the fear of raids, evictions, arrests, and jail. Then there is the material pressure. As squatters, many things are no longer possible. It has become very difficult to earn money which is urgently needed for the renovation and repair of houses that have been partially destroyed [...] Add to this the uncertainty in which we live. Everyday something happens (raids, demos, plenums) in which spontaneous decisions are necessary. And yet, we are still learning how to live in freedom and self-determination and that costs a lot of energy. We are all too often overcome by the new forms of our movement which we cannot handle.[67]

The repertoire of practices developed by squatters in West Berlin in the early 1980s thus generated intense feelings of joy and exhilaration and a sense that an another form of shared city life – autonomous and self-managed – was a distinct possibility. As new political horizons flickered into view, however, it became increasingly difficult for squatters to respond to apparent openings and defeats not to mention the intense media scrutiny they faced and the concomitant breakdown of solidarity within houses. Negotiations with local authorities were greeted by many activists with feelings of anger and betrayal, while evictions and police raids were accompanied by a heightened sense of loss, despair and hopelessness. For many squatters, evictions were tantamount to "the death of an acquaintance. It is the loss of something that has a key meaning (*Zentrale Bedeutung*) for you". Squatters understood their actions as 'world-making' while forced expulsions, they argued, destroyed the "relations between people" and brutally dismantled "their way of living".[68] If anger and indignation had been collectivising, despair individualised and foreclosed political alternatives and, as the autonomist journal, *Radikal* made clear, led to a widespread sense of "stagnation (*Stillstand*)" (see Gould, 2009: 396).[69] The squatting scene in West Berlin may have, in other words, reached its apogee in the spring of 1981. It had also begun to fracture and unravel.

From Resistance to Pacification: The Decline of the West Berlin Scene

In the summer of 1981, squatters associated with the Kreuzberg squatters' council (*Besetzerrat*) published a call-out for a radical festival of squatters and other like-minded activists. The TUWAT or ('Do Something') spectacle was modelled on

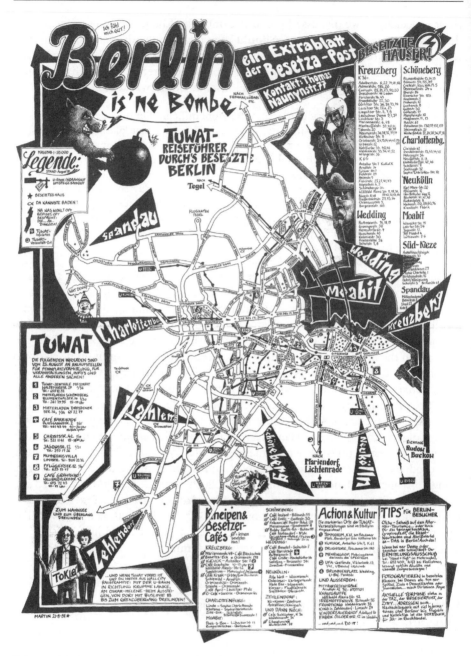

Figure 4.13 Map produced for TUWAT 'spectacle' in Kreuzberg, August 1981, *Instand-Besetzer-Post* (Papiertiger Archiv).

the TUNIX congress of 1978 and was designed to reanimate and revitalise a scene increasingly fractured and under pressure from local authorities. To coincide with the meeting, which was scheduled to begin on 25 August 1981, the *Instand-Besetzer-Post* published a special edition that included two large-scale maps (see Figures 4.13 and 4.14) that served as a guide to the squatting scene and the wider alternative infrastructure in West Berlin and Kreuzberg in particular. The first map of West Berlin included a list of all the squatted houses in the city that were represented, in turn, by a series of black dots. Cafés and bars associated with squatters were also included as well as meeting points for the TUWAT festival and other tips for visitors. On the flip side of the map, a detailed cartographic rendering of the Kreuzberg district was produced, zooming in on the location of squatted houses in the neighbourhood as well as key sites that were part of the TUWAT 'spectacle' (squatters' bars, cafés and information centres, bakeries and market halls). Additional sites within the local activist milieu were also identified (artisan workshops, bike repair shops, organic grocery stores, printing shops, secondhand shops, etc.).[70] Taken together, the maps provide a snapshot of the rich infrastructure that formed part of the alternative milieu that had emerged in West Berlin in the wake of the earlier TUNIX congress (see Scheer and Espert, 1982). If this was a scene that already counted over 100,000 members by the end of the 1970s, squatting only strengthened, so it seemed, the milieu's commitment to oppositional geographies that prefigured other forms of autonomous urban life.

In this context, the TUWAT meeting was ultimately seen as a failure, attracting only 3000–5000 attendees where 50,000 had been anticipated. For many, the meeting only contributed to the decline of a movement that had already begun to fracture in the spring of the same year. Whilst the dynamics of the scene since December 1980 were largely shaped by confrontation and antagonism aimed at exploiting the power vacuum that accompanied the 'Garski affair', there were many who welcomed a return to the strategy of negotiation and mediation adopted by citizens' action groups and tenants' representative offices. As Andrej Holm and Armin Kuhn have argued, tensions broke out between those who favoured further confrontation and others who advocated the "strategic pursuit of alternative urban political goals" including legalisation and the renovation of the houses they occupied. While squatters downplayed the conflicts between 'negotiators' and 'non-negotiators', it was clear that some were willing to hold on to houses at the expense of an earlier consensus that placed the release of 'political' prisoners and an overall solution for *all* houses as *the* condition for any negotiations (Holm and Kuhn, 2011: 647). In response, some of the 'non-negotiators' formed an autonomist faction that reverted to ever more militant tactics and strategies which anticipated the emergence of the *Autonomen* movement in the 1980s (Geronimo, 1992; A.G. Grauwacke, 2003).

The transitional Berlin government that formed in the wake of the Garski scandal in February 1981 adopted, in this context, an approach that facilitated the legalisation of some houses that met the prerequisites for immediate

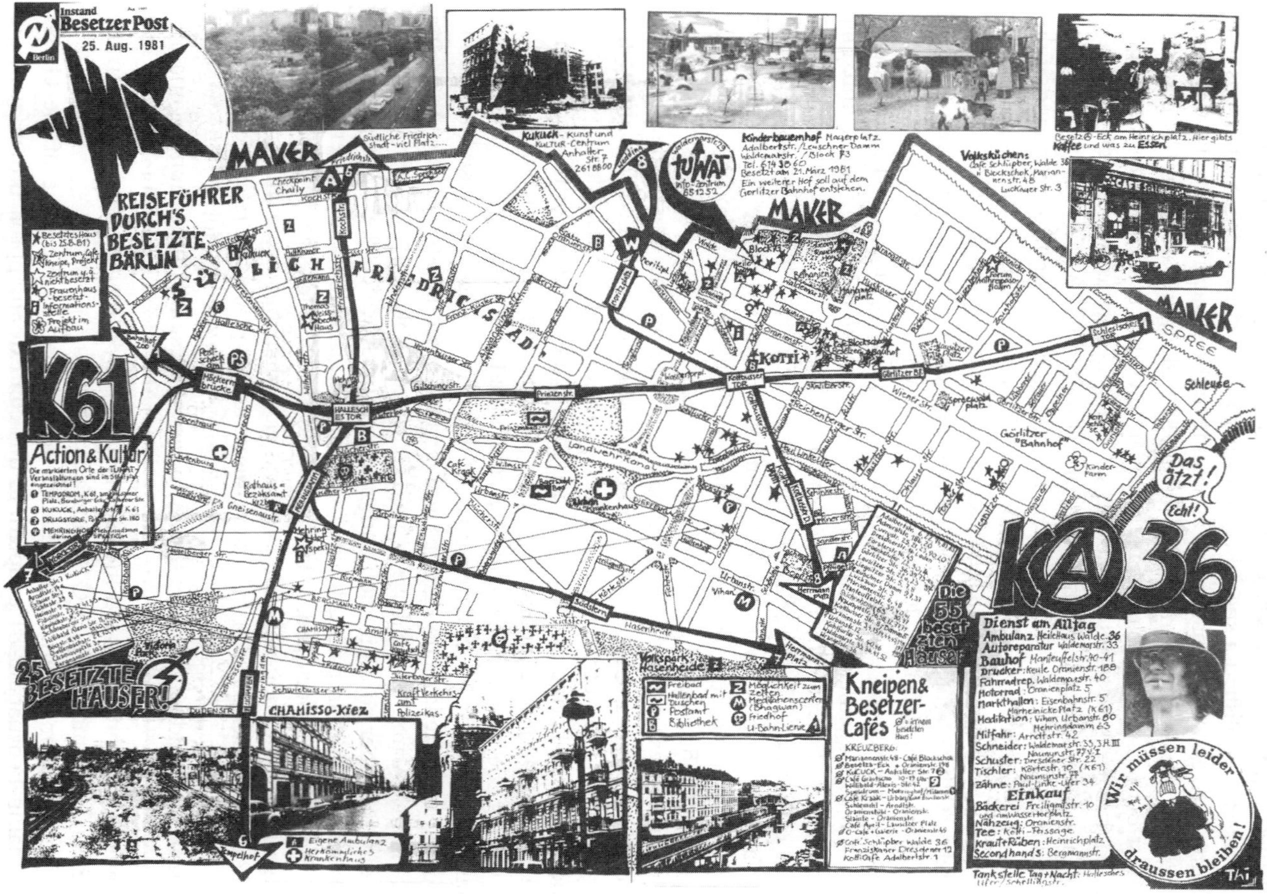

Figure 4.14 Second map produced for TUWAT 'spectacle' in Kreuzberg, August 1981. *Instand-Besetzer-Post* (Papiertiger Archiv).

'rehabilitation' though evictions and widespread police repression ultimately persisted (see Bodenschatz et al., 1983: 322). The defeat of the SPD in the May elections led to a new CDU-run Senate under Federal President Richard von Weizsäcker. While attempts were made by the Minister of Housing, Ulrich Rastemborski, to engage with the squatters, his counterpart in the Interior Ministry, Heinrich Lummer, chose a new hard-line policy that proscribed and vigorously policed any further attempts to squat in West Berlin. What became known as the 'Berlin Line of Reason' (*Berliner Linie der Vernunft*) was characterised by a large-scale offensive against existing squatted spaces and the further criminalisation of their occupants, while the city's boulevard press ratcheted up the rhetoric linking squatted spaces to earlier forms of leftist militancy and terror (Suttner, 2011: 203–204). The new policy was set into motion only a week into the new CDU government as a squat at Mittenwalderstraße 45 was cleared in a 5-hour police operation that involved over 1300 officers and 173 arrests (see Reichardt, 2014: 528–529).[71] The 'law and order' approach favoured by Lummer and many in the police culminated in a wave of searches, arrests and evictions that led to the clearance of eight squats on 22 September 1981 and the death of an 18-year-old squatter, Klaus-Jürgen Rattay, who was hit by a bus as he fled from the police (Reichardt, 2014: 531–532; Ermittlungsausschuss im Mehringhof, 1981a).

The death of Rattay marked, in many respects, the downfall of the West Berlin squatting scene. While efforts were made by some squatters and their supporters to found associations that would serve as intermediaries which facilitated the collective legalisation of houses (Netzbau and later Stattbau), they were repeatedly thwarted by the Interior Ministry who ordered evictions on the flimsiest of pretexts often during negotiations (Holm and Kuhn, 2011: 648; see Katz and Mayer, 1985: 35). By March 1984, only 18 squats remained. A few months later, the iconic KuKuCk (Kunst & KulturCentrum Kreuzberg) was raided by the police and on 8 November 1984, the occupants of the last squatted house in West Berlin were evicted. The strategy pursued by Lummer had, in other words, succeeded in breaking the squatting movement. Of the 169 squatted houses that were occupied in West Berlin at the high point of the squatting wave in May 1981, 105 had been legalised through a combination of rental and purchase agreements. The remaining houses had been cleared out (Holm and Kuhn, 2011: 648).

For many, such a programme of legalisation was tantamount to a form of 'pacification' that further divided the squatting scene and facilitated the criminalisation of hard-core autonomists. At the same time, legalisation also represented, as Margit Mayer has argued, an attempt to *institutionalise* the squatting scene through the assimilation of self-help practices by local authorities (2013b; see also Katz and Mayer, 1985). Many 'projects' that were able to guarantee long-term use of a building fell under the *Behutsame Stadterneuerung* programme later ratified by the Berlin House of Representatives in 1983. Under this programme, houses could apply for public funds to repair and modernise their properties through what became known as the 'Structural Self-Help' (*Bauliche Selbsthilfe*) initiative (Sonnewald and Raabe-Zimmerman, 1983).[72] At the same time, the International Building Exhibition (IBA), which had been set up in West Berlin in 1979, turned to the practices of local squatters as a template for a new model of urban redevelopment that favoured citizen participation, grassroots redevelopment and the self-management of existing housing. As Holm and Kuhn concluded, "the squatted houses not only triggered a new policy of urban renewal; they were at the same time a kind of experimental laboratory in which new instruments of urban renewal were trialled" (2011: 653).

The adoption by local authorities of a programme of cautious urban renewal thus marked a final break from an earlier era of top-down planning in West Berlin characterised by a particular emphasis on demolition and displacement. In the long run, it served, on the one hand, as a model for later rounds of urban regeneration in Berlin that sought to incorporate 'autonomy' and 'self-help' as a major mechanism in the commodification of urban space and which were predicated, in turn, on the re-appropriation of alternative forms of urban living and the further gentrification of existing neighbourhoods (Mayer, 2013a; see also Uitermark, 2004; Iveson, 2013). For many squatters, on the other hand, it also signalled a more pragmatic approach that demanded a trade-off between existing political institutions and new 'insurgent' forms of urban citizenship (Holston, 2008). It would be misleading, however, to conclude that the rights to a different city were ultimately contingent on a form of activism that attempted to balance and extend claims for self-determination with broader participation in local politics. To do so runs the risk of eliding the words and doings of squatters in West Berlin in all their messiness, vitality and emotional intensity. The various spaces and strategies, the different achievements and experiences, and the countless struggles and disagreements point to a vast *archive* of practices that promoted a different way of imagining the city which enjoyed significant popular support and found new form in the following decades as part of a growing alternative infrastructure that subsisted in a number of neighbourhoods, most notably Kreuzberg. Taken together, these orientations draw attention to the different ways in which new, provisional, often ephemeral and sometimes durable urban worlds are composed in settings of growing inequity (Simone, 2004: 240). For squatters, the question was not to simply demand housing but to *realise it for themselves* and take back and occupy the time and space that was refused to them and others in Berlin. While this depends on a critical perspective that draws attention to the sufferings and injustices of city life, it also re-casts the 'right to the city' as a 'right' to forge other different spaces (see Vasudevan, 2014a). If this chapter began with the words of Félix Guattari speaking at a 1978 congress in West Berlin, it concludes with those of Michel Foucault with whom Guattari shared the stage in Berlin that year. In a note for his last ever lecture, written only a few months before the final squatted house was cleared in West Berlin, Foucault concluded that "there is no establishment of the truth without an essential position of otherness; the truth is never the same; there can be truth only in the form of the other world and the other life". To produce a critical history and geography of squatting in West Berlin is to ultimately recognise and acknowledge the emergent possibilities of this "other world" and "other life" (Foucault, 2011: 340).

Notes

1 Papiertiger Archiv (hereafter PTA), flyer, "Treffen in TUNIX", n.d. Ordner TUNIX und Widerstand.

2 The description of West Germany as a 'security society' governed by a 'security state' is borrowed from a series of sketches made by the French philosopher Michel Foucault in 1977. Foucault was exploring the relationship between the deployment of exceptional extra-legal measures and the *securitisation* of the state (see discussion in Hannah, 2012).

3 Foucault and Guattari spoke together at the TUNIX meeting in a workshop session on "anti-psychiatry" (PTA, flyer, "Treffen in TUNIX", Ordner TUNIX und Widerstand). As Matthew Hannah (2012) has recently shown, Foucault's 1979 lectures on the German ordo-liberals trace the development of a specifically German mode of neo-liberal governmentality. Foucault began this project in 1977 and, as Hannah argues, was strongly influenced by the events of the German Autumn. The extra-legal dimensions of what Foucault called the 'security state' did not, however, appear in the 1979 lectures. According to Hannah, this was in no small part a product of a disagreement between Foucault and Gilles Deleuze over whether the West German state could be considered 'fascist'. If Foucault shied away from the implications of his earlier 1977 sketches, one could plausibly argue that Guattari's 1978 address at the TUNIX broadly followed and developed Foucault's initial argument.

4 *taz*, 25.1.2008.

5 *taz*, 25.1.2008. Interview with T.S. (Berlin, August 2010).

6 See also Hamburger Institut für Sozialforschung (hearafter HIS), "Italien: Kongress in Bologna, 23.–25. September", *Info-BUG*, 176, 10 Oct. 1977, 7.

7 The German fascination with the figure of the Native American Indian as a romanticised figure of resistance has a long and complicated history (see Sieg, 2002).

8 Brückner was himself a victim of the so-called Ban on Careers (*Berufsverbot*) which required all public servants – including academics – to uphold the West German *Grundgesetz* (Basic Law). A 1972 "Decree against radicals" issued by the Conference of State Interior Ministers required background checks of all applicants for state employment and bans for individuals who were or had been members of any organisations that were deemed to be 'anti-constitutional'. The decree led to lifetime bans for thousands of left-wing activists. Brückner himself was suspended from Hannover University for his activities (see Hannah, 2012: 120).

9 *taz*, 25.1.2008, 5.3.2008.

10 *Der Tagesspiegel*, 27.1.2008.

11 PTA, flyer, "Treffen in TUNIX", Ordner TUNIX und Widerstand.

12 PTA, flyer, "Treffen in TUNIX", Ordner TUNIX und Widerstand.

13 PTA, flyer, "Treffen in TUNIX", Ordner TUNIX und Widerstand.

14 The German feminist movement was largely absent from the meeting which only reinforced the marginalisation of women's issues with the wider leftist scene. Members of the 2. Juni Bewegung (June 2nd Movement) also criticised the inability of the meeting to overcome "the total fragmentation of the left". This fragmentation made it too easy, they argued, for the ruling class to "paralyse the dangerousness and appeal of the concentrated revolt in the streets" (quoted in Brown, 2013: 361).

15 HIS, "Reise zum Mond", *Pflasterstrand*, 28 (February 78) n.p.

16 PTA, flyer, "Stammtisch SO 36, Haberkern in Gefahr", Ordner Häuserkämpfe 1970s-8/1980.

17 Interview with T.S. (Berlin, August 2010).

18 By the early 1960s, the governing CDU had begun to lift rent control on older housing stock across cities in West German cities though Berlin was exempted from this programme.

19 PTA, Stadtteilzentrum Feuerwache Brochüre, December 1977, p. 3.

20 HIS, *InfoBUG*, "Tunix im Stadtteil", 11.1977, p. 11.

21 Kreuzberg Museum, hereafter KM, flyer, n.d., squatting boxfile.

22 There is no direct translation into English of "*Zweckentfremdungsverbots-Verordnung*" ("ZwVbVO"). It refers to a regulation prohibiting the misuse of housing, i.e. housing for purposes other than originally intended.

23 While the term "*instand(be)setzen*" is commonly translated into English as "rehab squatting", I have chosen, in this context, to translate the motto as "to squat and mend".

24 PTA, "Erklärung der BI SO 36", 1.24.1981, Ordner Häuserkämpfe, 1/1981; also available here: http://archiv.squat.net/berlin/12.12.80/1/IniSO36.html

25 PTA, *Südost Express*, Nr. 5/1979, p. 19.

26 *TIP Magazin* (June-July, 1979), p. 8–9.

27 PTA, *Südost Express*, Nr. 12/1979, p. 3; see also "Leuschnerdamm 9 – Instandbesetzt", n.d., Ordner Häuserkämpfe 1970s-8/1980.

28 PTA, *Südost Express*, Nr. 12/1979, p. 3.

29 PTA, Flyer, "Betroffenen-Vetretung, Dredener Straße Berlin 1979", n.d., Ordner Häuserkämpfe 1970s-8/1980.

30 PTA, flyer, "Besetzerrat-Info", n.d. Ordner Häuserkämpfe, 1970s-8/1980.

31 By the late 1970s, the West German police had already developed a series of protocols for evicting squatted houses. See Linker (1981) and PTA, "Grundkonzeption für vorbereitetes Handel bei Räumung von besetzten Häusern", uncatalogued copy of lecture at the Police Academy in Frankfurt, September 1977.

32 *Spandauer Volksblatt*, 30.5.1980

33 PTA, flyer, "Das neueste von der Chamisso 3", n.d., Ordner Häuserkämpfe, 1970s-1.1980; see also *Spandauer Volksblatt*, 5.6.1980; *Der Tagesspiegel*, 5.6.1980.

34 Quoted in *Spandauer Volksblatt*, 17.6.1980.

35 PTA, Besetzerrat, "Presseerklärung", 1.10.1980, Ordner Häuserkämpfe, 9-11/1980.

36 PTA, "BI SO 36 report", 14.12.1980, Ordner Häuserkämpfe, 12/1980; *Radikal Extrablatt*, December 1980.

37 *taz*, 15.12.1980.

38 PTA, "Bezirksamt von Kreuzberg, Presserklärung", 15.12.1980, Ordner Häuserkämpfe, 12/1980.

39 *Frankfurter Rundschau*, 14.12.1980; *Der Abend*, 15.12.1980.

40 *Der Tagesspiegel*, 16.12.80; *taz*, 22.12.1980.

41 PTA, "Runter mit der Besetzerwahn", unsigned letter dated 6.1.1981, Ordner Häuserkämpfe, 1/1981.

42 PTA, Besetzerrat, "Presseerklärung", 16.1.1981, Ordner Häuserkämpfe, 1/1981.

43 *taz*, 15.12.1980.

44 PTA, "Pressemitteilung Nr. 2 des Senators für Inneres – Pressereferent", 27.1.1982, Ordner Häuserkämpfe, 1/82-4/82.

45 PTA, pamphlet, "Hauskonzept – Manteuffelstraße 40/41", 10 October 1981, Ordner Häuserkämpfe, 10/1981.

46 PTA, flyer, "Nachrichten für Handwerk-kollektiv", January 1981, Ordner Häuserkämpfe, 1/1981.

47 PTA, *Instand-Besetzer-Post*, Volumes 1–5 especially. See, for example, "Strom: Einiges zur Elektrik", 11.3.1981, p. 6; "Neues vom Bauhaof: Wo findet man Strom", 17.3.1981, p. 10–11; "Wasser", 25.3.1981, p. 16–17; "Sein wir schlau am Bad", 1.4.1981, p. 16; "Sein wir schlau: irgendwo musse's raus – der Abfluss", 5: 18; "Kohlebadeofen: under Liebster", 9.4.1981, p. 18.

48 PTA, uncatalogued brochure, "Blockrevue, Operation Picobello, Block 101/103", p. 11.

49 PTA, pamphlet, "Hauskonzept – Manteuffelstraße 40/41", 10 October 1981, Ordner Häuserkämpfe, 10/1981.

50 Café Krantscho was located in a squat at Willibald-Alexis-Straße, Lokal Lummerland at Winterfeldstraße 38, the Besetzereck at Oranisenstraße 45, the Bobby Sands Pub at Bülowstraße 89 and Café Knüppel at Knobelsdorffstraße 40.

51 PTA, "Frauen besetzten Schokofabrik", *Instand-Besetzer-Post*, 15.5.81, p. 7.

52 PTA, "Betrifft: Besetzerinnen Kongress", *Instand-Besetzer-Post*, 19.6.81, p. 25.

53 PTA, flyer, "Hexenhaus. Wir haben dieses Haus besetzt – er wird von uns instandge-hext", 8 January 1981; Ordner Häuserkämpf, 1/81; see also *taz*, 7.1.81; *Abend*, 8.1.81.

54 PTA, Untitled flyer, n.d. Ordner Häuserkämpfe, 1/1981.

55 APO-Archiv, brochure "Frauercafé Moabit" (1982); PTA, "The Future is Female", *Instand-Besetzer-Post*, 12.6.81, p. 23.

56 PTA, "Türkische-deutsche Hausgemeinschaft, Kottbusser Str. 8", Ordner Häuserkämpfe, 7-28.2/1981.

57 PTA, "Tuntenhaus Presserklärung", December 1981, Ordner Häuserkämpfe, 11-12/1981.

58 *Die Zeit*, 12.08. 1983.

59 Freie Universität Berlin, "APO und Sozial Bewegungen" Archiv (hereafter APO-A), Landesamt für Verfassungsschutz beim Berliner Innensenator, "Der 'Häuserkampf' in Berlin (West)", p. 36, 38.

60 Another version of the same letter appeared in December 1981. See PTA, Schöneberger Besetzerrat, "Offenen Brief an die Bürger Berlins", 8.12.1981, Ordner Häuserkämpfe, 11/81 and 12/81; original emphasis.

61 Interviews with K.L. (February 2008), I.M. (August 2009, 2010) and C.L. (August 2009).

62 *taz-journal* 3, 1981, p. 106.

63 PTA, Der schwarze Kanal, "Ein Jahr Anarchie und Glück: Bericht über das Zusammenleben", *Der schwarzer Kanal*, 2.3.1982, p. 26.

64 *taz*, 7.4.1981.

65 PTA, "Zwischenbilanz der Instandbesetzer der Köpischerstr. 5", *Instand-Besetzer-Post*, 29.5.1981, p. 19.

66 PTA, "Vollversammliung, Besetzerrat, 7.4.1981, Ordner Häuserkämpfe. Reprinted as "Der Besetzerrat spricht", *Instand-Besetzer-Post* 7.4.1981, p. 3.

67 PTA, "...in Freiheit leben. Gedanken zur Verhandlungsfrage", *Instand-Besetzer-Post*, 22.5.1981, p. 4.

68 *taz*, 3.7.81; *taz*, 4.8.81.

69 PTA, "Stillstand ist das Ende von Bewegung", *Radikal* 6,1 (1982), pp. 12–14.

70 See PTA, *Instand-Besetzer-Post*, 25.8.81.

71 Landesarchiv Berlin (hereafter LAB) B Rep. 002, Nr. 16530, "Bericht zur Entwicklung und Beendigung des Hausbesetzerproblems". See *Der Spiegel*, 28.09.1981, pp. 26–32.

72 The Structural Self-Help programme was initiated in West Berlin in 1982 and offered public funds to legally registered nonprofit organisations and cooperatives in order to support DIY maintenance and repair. Until 2002, 80–85 percent of costs were subsidised for nonprofit builders. The remainder was to be obtained through "proprietary capital" and muscle mortage (*Muskelhypothek*). The programme was discontinued in 2002 (Heyden and Schaber, 2008: 142).

Chapter Five
Separation and Renewal

The house I live in is not very grand
So dark and drab, oh well, its 1910!
Such a wretched old house, falling apart.
Of which there are many in Prenzlauer Berg.
Reinhard Lakomy, "Das Haus wo ich wohne" (1974)

[The violence] was intended to show us our powerlessness. As we sat around a table [in the squat], it all turned pretty grim. But I knew: they could ruin our things and maybe they could hurt us physically. But they could do no better with their cops. The whole time I was glad I was sitting with everyone at the table. I was really happy that we were there together.
Eyewitness account by squatters of the Mainzer Straße evictions, 14 November 1990 (Arndt et al., 1992: 125)

The origins of this chapter can be traced back to the violent eviction of squatters from 12 houses on Mainzer Straße in the Berlin district of Friedrichshain on 14 November 1990 after three days of street battles with over 3000 West German police officers that resulted in the arrest of over 400 activists (Arndt et al., 1992; Amantine, 2012) (see Figure 5.1). The fall of the Berlin Wall a year earlier had marked the beginning of a rapid process of 'spatial redefinition' for the entire GDR. East German housing policy meant that much of East Berlin's 19th-century housing stock was never properly maintained and, by the late 1980s, had

Metropolitan Preoccupations: The Spatial Politics of Squatting in Berlin, First Edition.
Alexander Vasudevan.
© 2015 John Wiley & Sons, Ltd. Published 2015 by John Wiley & Sons, Ltd.

Figure 5.1   The eviction of squatters from Mainzer Straße in Friedrichshain, November 1990 (Umbruch Archiv).

slipped into serious disrepair (Sheridan, 2007: 102). With reunification, formerly nationalised property was seized and quickly transferred to private ownership while planning imperatives prioritised the urbanisation of capital and the creative destruction of previously socialist space (Heyden, 2008: 34). At the same time, the fall of the Wall offered a rare opportunity for various social groupings to create radically new and autonomous spaces of collective and common property. Just as a first wave of squatters had earlier moved into the West Berlin district of Kreuzberg in the late 1970s and early 1980s, a second wave now moved into abandoned tenement blocks in the east. At one point in 1990, over 130 buildings were occupied in various districts of East Berlin though the violent clearing of Mainzer Straße represented, in many respects, the beginning of the end of the squatter movement in Berlin and the transformation of the scene into a multitude of new experimental geographies (see Vasudevan, 2011a).

In this chapter, I examine the emergence of a second major wave of squatting in Berlin in the former East of the city. As a number of scholars have argued, the new wave of squatting that erupted in the winter of 1989 differed from earlier attempts to occupy empty apartments in the West and "can only be viewed within the context of the explosive social changes that took place during the turnaround (*Wende*) and reunification" (Holm and Kuhn, 2011: 649). If there is much to recommend in this view, especially when seen in the context of the enormous renewal require-ments that accompanied Berlin's attempt to re-imagine itself as a global city and the capital of a united Germany, my own aim is to shift some attention back to a more expansive history of knowledges, practices and networks that shaped the evolution of the squatting scene in East Berlin (Holm and Kuhn, 2011; see Colomb, 2011; Till, 2005). Who, after all, were these new squatters? Were they from the East or the West or both? Were there any squatters in East Berlin before 1989 and, if so, how did they challenge the dominant model of ownership and control in the GDR and promote an alternative vision of the city? And, finally, what were the central characteristics (goals, action repertoires, political influences) of urban squatting in the era after the fall of the Berlin Wall?

To begin to answer these questions demands a rethinking of the recent history of radical housing politics in Berlin. In this chapter, I build on the arguments set out in Chapters 3 and 4 in order to explore the complex combination of formal and informal practices – from planning, policy and law to everyday practices of dwelling and infrastructure – that shaped the development of squatting and other occupation-based practices in East Berlin before the fall of the Berlin Wall and in its immediate aftermath. The chapter shows how the large-scale squatting of buildings across East Berlin in 1989 and 1990 was, on the one hand, connected to an *earlier* history of illegal occupation and alternative housing in East Berlin and elsewhere in East Germany, a practice that was commonly known as *Schwarzwohnen* ('illegal living').[1] On the other hand, it also represented a further extension of the repertoire of contentious politics that played such an important role in the history of the squatting scene in West Berlin and would assume a new form in the early months of 1990 as a growing numbers of activists from the West moved into the East. If I set out in this chapter to retrace these oppositional geog-raphies, it is not my intention to argue that the recent history of squatting in dis-tricts of former East Berlin was somehow homologous to its earlier counterpart in the West. Nor do I seek to uncover an unbroken line of occupation and resis-tance linking the two major waves of squatting in Berlin. Rather, my aim is to examine and account for the full range of practices that shaped the dynamics of squatting in Berlin before and after the fall of the Wall and to situate the activities of squatters within an uneven landscape of dispossession and displacement, restructuring and regeneration. At stake here is a detailed reconstruction of an alternative urban imaginary during a period when the very terms of Berlin's iden-tity and status were up for grabs.

The chapter is divided into two main parts. The first part seeks to reconstruct the relatively unknown history of illegal occupation in East Berlin (*Schwarzwohnen*). Particular emphasis is placed here on the relationship between *Schwarzwohnen* and the articulation of alternative forms of domesticity and home-making that challenged official state priorities (see Grashoff, 2011a, 2011b; Vasudevan, 2013). Squatters in East Berlin were thus responsible for a spectrum of alternative prac-tices and tactics that spoke to the importance of housing for the development of oppositional cultures and networks in the German Democratic Republic (hereafter GDR) during the 1970s and 1980s. They also played, in this context, an important role in a new wave of occupations that erupted in the winter of 1989. The second part of the chapter tracks the further intensification of the squatting scene and its various material, emotional and political geographies. It also draws attention to the growing conflict between West and East German activists in the months leading up to official German reunification. Taken together, the chapter shows how the political and legal vacuum that accompanied the fall of the Berlin Wall allowed squatters to carve out new spaces of autonomy and experimentation that built on a rich repertoire of practices whose origins lie on *both* sides of the Wall. At the same time, it argues that these opportunities became increasingly limited in the face of the rapid restructuring of the housing market in the former GDR. The chapter therefore concludes by revisiting the series of events that led to the violent clearing of squatters on Mainzer Straße in November 1990 and the wider implications that the evictions had on housing-based activism in Berlin.

Schwarzwohnen: An Alternative History of Housing in East Berlin

In September 1988, an anonymous report appeared in the East German underground magazine *Umweltblätter* describing the plight of a group of squatters who had occupied Lychener Straße 61 in the Berlin district of Prenzlauer Berg. In the squatters' own words, they had "occupied the house in order to overcome the contradiction between, on the one hand, the many vacant and decaying houses [in Berlin], and, on the other, a growing number of people in search of housing". As "squatters (*Instandbesetzer*)," they proclaimed, "we will resist the further cultural and spiritual devastation of the country."[2] The *Umweltblätter* represented one of the most important and widely distributed Samisdat publications in East Germany during the 1980s.[3] It was published by the Umwelt-Bibliothek, an independent information centre that opened in the basement of the Zionskirche meeting hall in Berlin in September 1986. The Umwelt-Bibliothek was founded by a number of prominent environmental activists including Christian Halbrock, Carlo Jordan, Wolfgang Rüddenklau and Tom Sello and soon became a key site within a wider underground network of protest and dissent. *Umweltblätter* was edited by Rüddenklau and was the largest and longest-running dissident publication in the GDR (32 issues between 1986 and 1989, with a print run of 4000 per issue).[4] Whilst the *Umweltblätter* and other Samisdat publications focused on issues relating to the peace and environmental movement that had sprung up in East Germany in the late 1970s, a number of articles also drew attention to the illegal occupation of housing in cities such as Berlin (Figure 5.2).[5]

In recent years, a growing body of scholarship has emerged charting the evolution of oppositional protest cultures in the GDR from the late 1960s until the fall of the Wall (Glaeser, 2011; Kowalczuk, 2002; Kowalczuk et al., 2006; Moldt, 2005, 2008; Pfaff, 2006). The illegal occupation of empty apartments across East Berlin

Figure 5.2 *Umweltblätter*, cover of dissident underground Samisdat publication of the Kirche von Unten, East Berlin, 1988 (Robert-Havemann-Gesellschaft).

and elsewhere in East Germany has, however, received relatively little attention despite the fact that housing in the GDR was, as in the case of the West, highly politicised with the state assuming principal responsibility for its construction, maintenance and allocation (Fulbrook, 2005: 50; on the subject of *Schwarzwohnen*, see Grashoff, 2011a, 2011b). This is hardly surprising, perhaps, given the parlous state of housing in the wake of World War II. In the 1950s, central heating was available in less than 3 percent of residences. Only 30 percent of residences had a toilet and 22 percent a bath. Over 45 percent of housing in the GDR had been constructed before 1900 (Betts, 2010: 120). Administration was largely handled by the Communal Housing Association (*Kommunalen Wohnungsverwaltung* or KWV) which was responsible for 72 percent of East Berlin's properties. A further 15 percent belonged to housing cooperatives and the remaining 13 percent were owned privately and under the control of state-allocated trustees (Mitchell, n.d.: 2–3). Rents throughout the GDR were capped at 1930s levels and citizens were legally required to obtain permission from local housing authorities before they could take up residence at a given address and receive an official tenancy agreement (Grashoff, 2011b: 14–15).

By the early 1970s, official policy had only exacerbated the dereliction of older housing stock across East Germany and, despite an ambitious building programme undertaken during the Honecker era, an acute shortage of housing persisted, especially in inner city neighbourhoods as ideological priority was given to large, industrially manufactured estates on the outskirts of cities. There were still 600,000 people on official housing waiting lists in the early 1970s with waiting times averaging between six to eight years (Häusermann and Siebel, 1996: 169; Mitchell, n.d.: 7). The cost of demolition and construction could not, in turn, be met by the state and, as a result, thousands of properties fell into ruin and remained empty. As a secret report commissioned by the Socialist Unity Party of Germany (*Sozialistische Einheitspartei Deutschlands* or the SED) in 1985 noted, there were over 235,000 empty properties across the GDR with particular concentrations in major cities such as Berlin, Dresden and Leipzig. As late as 1990, the number had risen to 400,000 while over 89,000 families and 382,000 individuals were still without accommodation of their own (Buck, 2004: 344, 383; see also Häusermann and Siebel, 1996).

It is against a backdrop of persistent housing scarcity, that many citizens – often young but not exclusively so – chose to bypass the official allocation process and occupy properties illegally. What Udo Grashoff has recently described as "*Schwarzwohnen*" ("illegal living"), or "*wohnen in Abriss*" ("living in ruins") as some residents preferred to call it, can be traced back to the occupation of a small apartment in 1967 on Kleine Marktstraße in the East German city of Halle (Grashoff, 2011a). As Grashoff argues in the only existing study of squatting in East Germany, *Schwarzwohnen* was not a marginal phenomenon but involved thousands of citizens in the 1970s and 1980s in Berlin and other major cities including Halle, Dresden, Leipzig, Potsdam, Erfurt and Jena. Smaller towns also bore witness to the phenomenon. By the early 1970s, squatting was already widespread across East Germany and, in East Berlin, a series of official briefings were issued that drew attention to a rapid rise in "the illegal occupation of dwellings".[6] By 1983, a report put the number of people illegally occupying buildings in the Berlin district of Prenzlauer Berg alone at 800. A few years later, the number had risen to almost

1300 (Grashoff, 2011b: 19). Throughout the 1980s, it is estimated that there were at least 1000 cases per year in East Berlin.[7] The official statistics point to a scale of squatting that, if anything, rivalled its counterpart in the West.

Recent scholarly attempts to develop a typology of squatting in the context of capitalist urban development has yielded a range of approaches based on different goals and motives (see Pruijt, 2013; Vasudevan, 2014a). The history of illegal occupation in East Germany was similarly complex, and depended on a spectrum of practices that highlighted the various ways in which an alternative right to housing was articulated in the GDR. Many former squatters spoke of a uniquely East German phenomenon. *Schwarzwohnen*, so they believed, had far more to do with a desire to take control of their own lives and respond to basic housing needs (in most cases squatters did pay some form of rent). According to Grashoff, "political protest and publicly flaunted forms of otherness hardly played a role. People wanted to have a home of their own. Illegal occupation was an expression of individual self-help in the face of the State's inability to fulfill its own promises on adequate housing for all" (2011a: 5). As one former *Schwarzwohner* recollected, "we lived in squats and did not understand why such a fuss was made in Kreuzberg with banners and demonstrations. To occupy had little political resonance. Rather we needed the space... (*Das Besetzen hatte hier weniger eine politische Dimension, sondern man benötigte den Raum...*)" (Felsmann and Gröschner, 2012: 15). They were supported by another squatter who pointed out that, "[*Schwarzwohnen*] was not a political movement, we had no programme, no overarching concept, just a happy co-existence. All of it just happened [...] We did not squat houses, it was not a political act nor an act of aggression or an act of provocation. It was simply a matter of course, because there was free living space and so we took it. That was '*Schwarzwohnen*' – a typical East German practice that no longer exists today" (quoted in Grashoff, 2011b: 11).

In the case of many squatters, the practice of occupation thus chimed with what German historians of everyday life (*Alltagsgeschichte*) have elsewhere described as "*Eigensinn*" as a means to account for the varied micro-practices through which "people have 'appropriated' – while simultaneously transforming – their world" (Lüdtke, 1995: 7). The concept of "*Eigensinn*" is commonly defined as self-reliance, self-will or the "act of re-appropriating social relations" and has become a key theoretical placeholder both within and beyond modern German history (Eley, 1989: 323–324). As Kathleen Canning has argued, "*Eigensinn* encompasses both historical subjects' encounter with 'constraints and pressures' and their [re]-appropriation [...] of these structural or discursive pressures" (2006: 105). For Paul Betts, it was in fact the private sphere and the everyday geographies of the home that, in the GDR, became an arena of autonomy, self-expression and potential dissent. "State socialism," he writes, "may have severely curtailed the limits of permissible talk, patrolling the borders of public discourse as much as it policed its state borders." "Nonetheless, East Germans," he continued, "were by no means silent. They grumbled and groused and sometimes resisted, and used the pressing problems of their private life not simply to defend a private sphere [...] In this sense, the private sphere was less a zone of immunity than a social assertion and even political claim" (2010: 15).

If *Schwarzwohnen* was one example of how citizens found a way to carve out an alternative personal space, many squatters were also connected to a wider network of opposition and dissent and saw their actions in terms that were constitutively political.[8] Wolfgang Rüddenklau, a former occupant of a squatted house in Prenzlauer Berg at Fehrbelliner Straße 7 explained: "The islands of occupied flats and houses [...] grew together to form an alternative social structure. They affirmed a self-determined lifestyle and developed a common culture" (quoted in Moldt, 2005: 7). Tina Bara, another former resident of the same house, described her four years in the Fehrbellinerplatz as "intense despite chaotic circumstances". "A pretty wild mixed bunch of people lived there," she added, "our apartment was a meeting place [...] especially for those in the illegal environmental movement. People constantly came to see us. Appointments and plans were made and projects forged. Many stayed overnight [...] There was a lot of discussions and drinking, and there was always something to celebrate. We were, of course, spied on" (Felsmann and Gröschner, 2012: 40). The connection between underground political activism and *Schwarzwohnen* was also echoed by Jörg Zickler, a former member of the Kirche von Unten (KvU), an oppositional movement in the GDR that was closely connected to the Protestant Church. As he later noted, "from my own political understanding, [*Schwarzwohnen*] was far more than an attempt to simply occupy an apartment. Rather it was an attempt to create a free space for people, against or parallel to the state. In my view, 'squatting' was the more appropriate word" (quoted in Grashoff, 2011b: 11).

It was clear therefore that a number of political activists did establish themselves in illegally occupied flats and houses and that an action repertoire more commonly associated with the West had been adapted and re-imagined by dissident groups in the GDR (Mitchell, n.d.: 5).[9] The occupied house on Fehrbelliner Straße was widely known to security services as a "non-conformist centre for resistance" while the squat at Lychener Straße 61 was home to eight different groups (Grashoff, 2011b: 152).[10] Both houses played an important role in the alternative infrastructure – social, cultural and political – that emerged in East Berlin in the 1970s and 1980s and which was centred on the district of Prenzlauer Berg. While dissident groups were scattered across the city, it was the neighbourhood with the highest concentration of groups involved in "state-independent political activities". This included groups such as 'Women for Peace' ('*Frauen für den Frieden*'), 'The Initiative for Peace and Human Rights' ('*Initiative für Frieden und Menschenrechte*') and 'The Environmental Library' ('*Umwelt-Bibliothek*'). It was, in turn, dissidents based in Prenzlauer Berg who were responsible for the underground publication and distribution of the most important Samisdat publications in the GDR such as *Grenzfall* and the *Umweltblätter* (Glaeser, 2011: 345–346). If Kreuzberg had established itself in the 1970s as a key site within the anti-authoritarian Left in the Western half of the city, the political scene in Prenzlauer Berg was "arguably [East] Berlin's and perhaps even the GDR's most important nodal point for non-SED dissident activities" (Glaeser, 2011: 346).

East German activists and dissidents were nevertheless mindful of the circumscribed sphere of action in which they moved. Public acts of squatting were largely eschewed as squatters adopted a more covert form of occupation which is why they often referred to their own activities as "*stilles Besetzen*" ("covert squatting"). Even the most political of squatters tended to locate their activism *outside* of houses and *inside* other more secure spaces. The Protestant Church, in particular, provided a protective infrastructure for a growing net of dissident practices. At the same time, many citizens in the GDR were still willing to accept the risks that accompanied squatting, as it was increasingly viewed as the only viable option for obtaining a home. While official channels proved laborious, there was no shortage of empty housing in major cities in the GDR. It was even reported in an article in the *Instand-Besetzer-Post*, the West German squatter magazine, that one could easily find "a dozen empty flats in Prenzlauer Berg" on any given afternoon.[11] As one former squatter recollected, all that was needed was a simple skeleton key and a bit of luck. "With some experience," they added, "you could pull it off without any hassle or disturbance. Once you were in, you were in. And then you have to quickly [...] register with the authorities and transfer rent in order to legalise everything" (quoted in Grashoff, 2011b: 101; see also Felsmann and Gröschner, 2012: 432). Many other squatters also believed that they simply needed to pay rent and utilities (usually for three to six months) to formalise their status. "After six months you'd go to the Communal Housing Association," explained one former squatter, the poet Peter Wawerzinek. "Normally," he continued, "you'd receive no more than a fine, and then you were safe and secure in your apartment" (Felsmann and Gröschner, 2012: 366).

And yet, squatting remained illegal in the GDR. State officials were reluctant, however, to carry out evictions as squatters often took pressure off growing waiting lists and, in many cases, returned dilapidated housing stock into use. As the right to housing was enshrined within the GDR's 1949 constitution, authorities also tended to avoid forced evictions as they would otherwise be legally required to provide alternative housing for squatters, many of whom had no other options.[12] In response, authorities often resorted to fines ranging from 50–500 M. There is insufficient evidence to gauge how many *Schwarzwohner* were ultimately fined although squatters in Berlin found that local authorities were more likely to issue penalty charges in Prenzlauer Berg than in the neighbouring district of Friedrichshain (Grashoff, 2011b: 101).[13] Unsurprisingly, perhaps, the files of housing authorities across the GDR were full of denunciations by neighbours who contacted the authorities to inform them of illegal occupations though the chances of remaining in properties remained high.

Squatting in the GDR was, in this way, characterised by a complex process of adaptation and negotiation that operated at the margins of legality and formality (see Holston, 2008; Datta, 2012). On the one hand, local authorities often turned a blind eye to occupations and adopted a pragmatic approach that balanced legal imperatives with immediate housing needs. On the other hand, *Schwarzwohnen* remained a fraught process that depended on a certain degree of acceptance by neighbours and the wider community. By the early 1970s, East German authorities had significantly enlarged the 'Housebook' (*Hausbuch*) programme so that it covered over half of all East German residential communities and 75 percent of East Berlin housing estates (Betts, 2008: 121). The *Hausbuch* was a register that included the name, birthdate and occupation of all tenants in a given house. Visitors from within the GDR who were staying longer than three days were required to report to the person in charge of the *Hausbuch* (known as the 'Residential Confidence Man' or *Hausvertrauensmann* [sic]) while visitors from outside of the GDR had 24 hours to do so (see Betts, 2010: 28). A number of former squatters recalled how the *Hausbuch* system did, in some cases, pose a challenge though they also found that many neighbours were helpful and offered assistance and advice (Grashoff, 2011b: 71). If anything, the rise of *Schwarzwohnen* in the 1970s and 1980s was part of a growing body of *informal* practices and tactics used by citizens in East Germany in response to housing insecurity and scarcity (subletting, house swapping, bargaining with authorities, etc).[14]

The properties occupied by squatters often required significant renovation. When Tina Bara and others squatted a house at Fehrbelliner Straße 7, they found that the "roof was broken and the apartment block had been cleared by housing officers." "We had chosen," she added,

> The summer holidays, since the authorities were, according to our own experience, a bit sluggish and not very efficient [...] It was a while before someone in the local housing office found out that we had moved in. As punishment, we were not able to secure a rental agreement. Rather we obtained a special license that stated that the authorities were no longer responsible for the apartment and that we had to do all the repairs ourselves. That summer it rained incessantly, the stucco fell from the ceiling, and in one room, a supporting beam came down. We tried our best to do makeshift repairs. When winter came, the water line was often frozen in the kitchen and we had to get the water from a tap in the loo. There was no bathroom and our attempts at heating were adventurous to say the least (quoted in Felsmann and Gröschner, 2012: 39).

Others squatters reported similar experiences. Ulrike Poppe, one of the founders of the 'Women for Peace' ('*Frauen für den Frieden*') group, recalled living in a ground-floor flat on Wilhelm-Pieck-Straße: "it consisted of one room, a kitchen and an outside toilet as was the case for many. Three walls were covered in damp, and there was as good as no natural light" (Felsmann and Gröschner, 2012: 306). The journalist and writer, Annete Gröschner, described her first flat on Schönhauser Allee as a dark 'cave'. "There was a kitchen," she noted, "with an original sink that was covered in a brown-green film. The floorboards were rotting and the oven had rusted through. But the worst part was the outdoor loo. The door had fallen off [...] and I had to use the now demolished toilet at the U-Bahn stop" (Felsmann and Gröschner, 2012: 433). As another squatter in a special report in the *Instand-Besetzer-Post* on the East Berlin squatting scene concluded, "without a bottle of schnapps, you would freeze your ass off."[15]

It is perhaps not surprising that, like their counterparts in the West, *Schwarzwohner* in the GDR adopted a *makeshift urbanism* that often relied on similar do-it-yourself practices that gradually transformed squatted flats into liveable spaces and that were, in turn, a product of a wider shadow economy based on practices of informality and reciprocity. Floorboards and pipes were replaced, roofs were repaired, and courtyards were transformed into small community playgrounds (see Grashoff, 2011b: 76–83). While a preoccupation with mending and repair highlighted the degree to which squatters re-imagined the home as a site of individual freedom and self-determination, re-appropriation and renovation were also mobilised as negotiating tactics with local authorities. In some cases, housing officers were willing to grant temporary licenses to squatters so long as they agreed to undertake and finance repairs. In other cases, authorities tacitly accepted the existence of illegal occupations, arguing that they ensured that buildings were occupied and in relatively good condition. One student thus recounted how he 'squatted' a dilapidated apartment on Belforter Straße in Prenzlauer Berg which had been boarded up by local authorities on hygienic grounds. He cleaned up and renovated the space, drying the damp walls with an electric heater before inviting housing officials to inspect his work at which point he was promised an official tenancy agreement.[16]

The everyday geographies of renovation and repair should not, however, detract from the political imperatives that played a growing role in the action repertoire adopted by East German squatters in the 1980s. In the early 1980s, a flat at Mühsamstraße 63 in Berlin-Friedrichshain was occupied and transformed into a 'commune for children' (*Kinderkommune*) by a group of adults and four children. The commune survived for three years until personal conflicts led to its dissolution (Grashoff, 2011b: 114). An earlier self-organised *Kinderladen* that was founded by Ulrike Poppe in 1980 in a ground-floor apartment on Husemannstraße in Prenzlauer Berg also existed for three years until it was forcibly cleared. Another nearby house on Dunckerstraße was well-known to state security organs as a key site of dissent and included a number of activists who were involved in practices that connected local environmental issues to a broader struggle to re-imagine city life as part of a shared political project (Halbrock, 2004: 108; Grashoff, 2011b: 142). When the house was renovated, the residents received rental agreements and were split up though many ended up moving illegally into a house on Lychener Straße (no. 61) which became a new centre for an underground scene which, by the late 1980s, also counted a number of illegal bars, cafés and galleries.

While the security services were, in this context, relatively well informed on the practice of *Schwarzwohnen*, they tended to limit their own activities to houses that were associated with dissident political groups. The occupants of Lychener Straße 61, for example, were subject to a number of police raids. The eventual eviction and demolition of the house in September 1988 was greeted by an unusually open display of defiance by the squatters who dropped a banner from the roof proclaiming that "this house has been ruined by the KWV". Additional banners were added with the words "rebellion, resistance, Lychener is fully in our hands" and "we might be out, but we will continue".[17] If these words seem prophetic, it is worth noting that, for most *Schwarzwohner*, occupation was still viewed less as an act of official political resistance and rather as an attempt to articulate a right to housing that saw the home as a locus of autonomy and freedom. In the end, squatters never posed a serious threat to the SED and its apparatus of control. And yet, it is nevertheless clear that the spectrum of alternative practices and tactics developed by squatters in East Berlin spoke to the importance of housing and the city for the development of oppositional cultures and infrastructures in the GDR. It was these practices – informal, makeshift and precarious – that laid the groundwork for the new wave of squatting that erupted in the Eastern half of the city in the winter of 1989.

Occupying a Vacuum: Squatting after the Fall of the Berlin Wall

The fall of the Berlin Wall in November 1989 and the rapid dismantling of the East German state apparatus in the months that followed have come to represent one of the defining moments in recent German history. While, by the early 1980s, the GDR's system of coercive surveillance had contributed to the emergence of an underground network of protest and dissent centred on Church involvement and a nascent peace and environment movement, it was the dramatic explosion of large-scale collective action in 1988 and 1989 that ultimately undermined the regime's stability (Pfaff, 2006: 12). Recent scholarship has, in this context, drawn attention to the failure of opposition activists to exploit the political vacuum that accompanied the collapse of the GDR (Pfaff, 2006; Rucht, 1996). Many members of the civic opposition movement were unable to adapt to the widening rebellion, choosing reform over revolution. And yet, as Andrej Holm and Armin Kuhn have argued, it was "the political power vacuum of the *Wende* period, and the massive loss of authority on the part of the police and municipality" that played a crucial role in the large-scale occupation of vacant old buildings in East Berlin (2011: 649). During this period, over 130 houses were occupied by squatters in the inner-city districts of Mitte, Prenzlauer Berg, Friedrichshain and Lichtenberg (Amantine, 2012).[18]

What is widely seen as the second major wave of squatting in Berlin has its immediate origins in the covert occupation of houses in the months before the fall of the Wall. As I have already argued in this chapter, during the 1970s and 1980s, squatters across East Germany developed an action repertoire that *anticipated* the large-scale occupation of buildings across East Berlin in 1989 and 1990. The practice of *Schwarzwohnen* continued right until the collapse of the GDR. In the spring of 1989, for example, a series of flats in a backyard apartment block on Prenzlauer Allee were illegally occupied by a group of *Schwarzwohner* who only made public their activities as 'squatters' in January 1990. It was, however, another nearby house in Prenzlauer Berg at Schönhauser Allee 20/21 that became the first 'official' squat of a new movement. While the house was first occupied in August 1989, it was on 22 December that the occupants chose to publicise their actions by dropping banners from the roof. As they later wrote in an article in *Interim*, an important publication in the West Berlin 'autonomous' scene, "this house is first and foremost our livelihood though we hope that it also serves as a model for self-determined collective living". "We have decided," the article continued, "to use our own labour and resources to repair and shape our living space (*Wir haben beschlossen, unseren Wohnraum mit eigener Kraft wieder Instand zusetzen und zu gestalten*). In order to realise this project, we will need the solidarity of the people." At stake here, as a former activist recalled, was "the idea that you can occupy a house as a group and that [radical] politics was no longer confined to the living room as was the case for all those years during the GDR. One now had the free space to act on the outside". The house thus opened a number of different initiatives and quickly became a key site within the new activist milieu that emerged during the *Wende* period.[19]

According to Andrej Holm and Armin Kuhn (2011), the new wave of squatting in the former East was characterised by three main phases; a periodisation supported by first-hand accounts, interviews and other contemporary sources. The first phase, from December 1989 to April 1990, was largely confined to the districts of Mitte and Prenzlauer Berg where over 70 houses were eventually squatted (Holm and Kuhn, 2011: 650). As in the case of Schönhauser Allee 20/21, houses were squatted openly with banners, secured windows and barricaded doors to protect the occupants from attacks by groups of neo-Nazis who had begun to organise in large numbers in the final years of the GDR and were themselves occupying a series of houses on Weitlingstraße in Berlin-Lichtenberg (Holm and Kuhn, 2011: 650; Brand and Fregonese, 2013).[20] A 'right to the city' was, in this way, actively contested in the months and years after the fall of the Wall though many squatters increasingly focused their attention on defensive measures at the expensive of wider anti-fascist activism.[21] The first phase was largely driven by East German squatters who were already connected to a range of alternative political organisations and subcultures, including citizens' action groups, who were challenging the planned demolition of countless housing blocks across Mitte and Prenzlauer Berg. Within a couple of months, there were 20 occupied houses across East Berlin and, in January 1990, the first squatters arrived from the Western half of the city. While houses at Kastanienallee 85/86 in Prenzlauer Berg and Köpenicker Straße 137 in Mitte were heralded as symbolic sites of cooperation between activists from the East and West, it wasn't long before squatters from the West outnumbered their Eastern counterparts.[22] By February 1990, no new houses were occupied by East German squatters. Occupations at Brunnenstraße 7, Tucholskystraße 32 and Linienstraße 206 were entirely made up of activists from the West and elsewhere.[23] At the same time, a number of the new squats including the 'Art Department Store' at Oranienburgerstraße (also known as 'Tacheles'), the 'Autonome Aktion Wydoks' on Schönhauser Allee and the 'Eimer' at Rosenthaler Straße placed particular emphasis on underground artistic activities and established themselves as a vital part of the wider 'scene' that in the early 1990s transformed the former East into a laboratory of autonomy, creativity and experimentation (see Gutmair, 2013).

The second phase of squatting lasted from the end of April 1990 to July 1990 and shifted to the district of Friedrichshain. Holm and Kuhn describe a "qualitative and quantitative expansion" as the number of squats rose to over 130 (2011: 651). It is in this very context that a writer and activist with the journal *Telegraph*, the successor to the East German Samisdat publication *Umweltblätter*, received confidential information about housing speculation and regeneration plans that linked the KWV in Friedrichshain with a property developer in West Berlin. The plans provided a veritable blueprint for the creative destruction of the neighbourhood as large parts of the district were earmarked for acquisition and joint management including houses on Niederbarnimstraße, Mainzer Straße and Kreutzigerstraße.[24] It was the role of the KWV, in particular, to generate conditions that would ensure lucrative profits for the developer. According to the proposed plans, the KWV was responsible for renovating flats according to standards designed by the developer. Ground floors and first floors were to be renovated to accepted Western norms while the remaining floors were to follow the cheapest GDR specifications.[25]

In response to these plans, the April 1990 issue of *Interim* included a piece by a small group of squatters already occupying houses in Friedrichshain as well as other activists from the East and West. The article drew attention to houses in Mainzer Straße that had been left vacant since 1987, and put out a call to "occupy these houses before it was too late" (see also Arndt et al., 1992).[26] In the following weeks, the occupation of abandoned houses in the East intensified and, on 30 April 1990, 11 vacant houses on Mainzer Straße were squatted by activists mainly from the West Berlin autonomous scene. Houses 2-11 were the first to be occupied though houses on the other side of the street including numbers 22/24 were soon squatted as well (Arndt et al., 1992; Holm and Kuhn, 2011).[27] The 'Mainzer' became the centre of the Friedrichshain squatting scene and was characterised by a range of political activities and housing projects: no. 3 was a woman's-only squat with a café; no. 4 housed the latest incarnation of the 'Queenhouse' (*Tuntenhaus*) and an antiquarian bookstore ('the Max-Hoelz-Antiquariat für DDR-Literatur')

Figure 5.3   The occupation of abandoned apartments in Mainzer Straße by squatters in 1990 (Umbruch Archiv).

on the ground floor; no. 5 was the home of the neighbourhood information centre (or 'Infoladen') and no. 6 opened a popular pub; there was a late night shop in no. 7 and no. 9 was responsible for a community kitchen which was open to the public; another communal kitchen and cinema would later open in no. 22 (see Arndt et al., 1992: 43–55).[28] In the end, over 250 squatters made Mainzer Straße their home in the summer of 1990 (see Bashore, 1991, 1992) (Figure 5.3).[29]

The third and final phase of squatting began in July 1990 and lasted until the violent eviction of the Mainzer Straße squatters in November of the same year. During this period, the number of houses dropped as authorities in East Berlin adopted their own version of the *Berliner Linie* (Holm and Kuhn, 2011: 651). While squatters across the Eastern half of the city had already established a council (the *Besetzerrat* or *B-Rat*) in January 1990 in the offices of the Kirche von Unten, meetings were infrequent and largely devoted to the organisation and coordination of resistance against attacks perpetrated by fascist groups and football hooligans. It was only in June that a concerted effort was made to transform the *B-Rat* into a political organisation. By this time, a small number of houses had been able to secure contracts in separate negotiations with local KWVs who were themselves undergoing a process of transformation and 'Westernisation' which would see them become limited liability companies. The status of any agreements made during this period was, however, nebulous at best and a decision was made by the *B-Rat* on 22 June 1990 to form a 'Contract Committee' or *Vertragsgremium* in order to negotiate directly with the East Berlin City Council (hereafter Magistrat). As the Committee made clear in a letter to the Magistrat dated 26 June 1990, "we urge the Berlin City Council to express a clear political commitment to the negotiation of contracts for all represented houses". The letter concluding by asking the Magistrat to issue a statement that guaranteed that no houses would be cleared and that an earlier contract with the neo-Nazi occupants of a house on Weitlingstraße would be terminated.[30]

The first meeting between squatters and Berlin Magistrat was held on 27 June and was an unmitigated disaster as the representatives of the Magistrat were neither qualified nor prepared to negotiate with the 'Contract Committee'. A follow-up meeting took place on 4 July. This time the Magistrat was represented by a team that included an expert consultant from the West, Hugo Holzinger, who had previous experience working with squatters in Kreuzberg during the early 1980s (see Rada, 1991).[31] In the meeting on 4 July, Holzinger offered the squatters an updated version of the trusteeship model that had been used in West Berlin in conjunction with intermediary organisations such as Stattbau, SPI and Wohnstatt.[32] The squatters countered with a suggestion that they would found their own trusteeship which would assume responsibility for the legalisation of all squatted houses and their renovation. As Holzinger later recalled in an interview with the *tageszeitung*, there was little appetite amongst city officials and politicians for a collective settlement with the houses represented by the 'Contract Committee'. If anything, the opposite was true as Holzinger quietly commissioned the West German law firm, Knauthe & Partner, to investigate whether there was a case to be made in East German law for evicting the squatters.[33] At the same time, and as a letter from the Berlin Magistrat dated 24 July 1990 also made clear, the further occupation of houses and other spaces was to be vigorously proscribed as municipal authorities applied the *Berliner Linie* ordinance to districts across the Eastern half of the city. New squats would no longer be tolerated and would be subject to eviction by the police within 24 hours of occupation.[34]

In the months that followed, negotiations stalled between the squatters and the Berlin Magistrat. While city officials attempted to negotiate with individual houses in a bid to foster divisions within the wider squatting scene, it was existing tensions within the *B-Rat* and internal disagreements between activists from the West and the East that finally came to a head in the fall of 1990 (Arndt et al., 1992).[35] Such a fracturing of solidarities was soon reinforced by a new round of evictions that brought the third and final phase of squatting in East Berlin to a violent end (Holm and Kuhn, 2011: 651). With official German reunification in October 1990, it became clear for many that the Berlin Magistrat was simply buying time until the problem of squatting would come under the jurisdiction of authorities from the West.[36] There were even unsubstantiated rumours that West Berlin police officers were using a ghost town on the outskirts of the city to train for the clearance of the Mainzer Straße squatters. It was activists living in two houses in Prenzlauer Berg and Lichtenberg, however, who were the first to be forcibly evicted by the police on 12 November 1990. A solidarity demonstration was quickly organised by a group of squatters living on Mainzer Straße though they were only able to reach Frankfurter Allee before running into a wall of police officers equipped with water cannons, armoured personnel carriers and stun grenades. The confrontation soon escalated as the squatters responded to the use of water cannons and tear gas with stones, flares and Molotov cocktails (Arndt et al., 1992: 13).[37] The battle between squatters and the police raged until the early hours of the morning. Over 1500 officers were deployed as the police tried and ultimately failed to force their way onto Mainzer Straße (Holm and Kuhn, 2011: 651). The squatters strengthened their barricades in anticipation of another police assault. As they reported, "an architect told us how deep we must dig our trenches so that no more bulldozers come through while local construction workers showed us how to use a jackhammer and an excavator. A neighbour installed a loudspeaker in his window so that he could direct us in the construction of the barricades."[38] In the end and despite efforts by local politicians to broker a peaceful solution, the West Berlin police used the violent resistance offered by squatters as a pretext to intensify their efforts and clear all of the occupied houses on Mainzer Straße. On 14 November 1990, an operation involving 3000 officers, 10 water cannons and a squad of helicopters was finally successful (Figure 5.4). Over 400 activists were arrested and there were countless casualties on both sides. The sheer scale of the violence involved was vividly captured in an account by one of the squatters involved. "The whole neighbourhood," they recalled,

...was under a shroud of tear gas I kept a cloth to my face and I walked down Gabelsbergerstraße because everything else was already cordoned off by the police who were standing there with their gas masks. It was hard to see through the fog. I was able to endure it for five minutes then I had to leave. I went back only to stop in a doorway to catch some air [...] I then passed through a backyard onto Frankfurter Allee. The cops were already in the Mainzer [Mainzer Straße]. I could see how the people looked as they were taken away by the police: exhausted, unable to walk, torn and beaten. They looked battered, some were bleeding, others had head injuries, the wounds had not been bandaged (quoted in Arndt et al., 1992: 112).

The clearance of the 'Mainzer' marked a violent turning point in the wider history of squatting in Berlin and led to the eventual fragmentation of the scene into a complex archipelago of accommodation, resistance and experimentation (Vasudevan, 2011a). As I describe in detail in Chapter 6, few squatters were willing in the aftermath of the Mainzer Straße evictions to reject a negotiated settlement with local officials. Over 75 percent of remaining occupied houses were contractually safeguarded as squatters turned to structural improvements and repair work, often under the auspices of public funding programmes (Holm and

Figure 5.4   During the Mainzer Straße evictions in Friedrichshain, November 1990 (Umbruch Archiv).

Kuhn, 2011: 651). The Berlin Senate spent over 300 million Euros in the 1990s as part of a new round of 'self-help housing' that led to the renovation of over 3000 units, many of which were in formerly squatted houses. In many respects, there-fore – and not wholly unlike an earlier history of occupation and activism in West Berlin – the wave of squatting that erupted in the final months of 1989 played an important role in the rapid regeneration and transformation of Berlin's inner-city neighbourhoods.

And yet it would be misleading to see squatters as the key engine of creative destruction during the *Wende*. Whilst attempts were made to co-opt and incor-porate elements of the squatter movement into an image of a new Berlin – creative, edgy and experimental – the everyday practices deployed by squatters also testified to the articulation of an alternative right to a city that had recently experienced a period of profound spatial and temporal upheaval. As Gisa Weszkalynys reminds us, institutions, infrastructures and imaginations were all dramatically altered in the years after unification (2010: 15). Weszkalynys is sup-ported by a number of other scholars who have shown how space and geography, more generally, played a central role in the formation of new identities as 'East' and 'West' were "reinvented not as innocent geographical reference points but as categories of description and self-description, conveying ideas about difference in [a]... united Germany" (Weszkalynys, 2010: 14; see also Berdahl, 1999; Borneman, 1992; Glaeser, 2000; Till, 2005). It was, hardly surprising, in this context, that East and West Berliners were "engaged in interpreting and reinter-preting the spaces in which they live" (Glaeser, 2000: 45). The city, after all, had become a site of change, confusion and disorientation.

For many commentators, it was the 'new' Berlin's "emptiness" that became the dominating spatial feature of the city. "The notion of Berlin as a void," writes Andreas Huyssen, "is more than a metaphor, and not just a transitory condition" (1997: 62; see also Doron, 2000; Oswalt and Stegers, 2000). As Huyssen argues, the creation of void spaces has played a central role in the history of Berlin as a modern metropolis and is constitutive of the city's complicated and ongoing rela-tionship with the politics of collective memory. For planners, architects and poli-cymakers, Berlin's numerous empty spaces, abandoned buildings and derelict industrial sites offered, in turn, an opportunity to re-imagine the city as the capital of a unified nation and as a post-socialist world city (Cochrane, 1999; Colomb, 2012; Till, 2005; Ward, 2011; Weszkalynys, 2010). And for yet others, abandoned apartments, vacant commercial spaces and plots of disputed wasteland became sites of possibility and experimentation, gathering and subversion. Whilst devel-opers and planners devoted their efforts to "filling Berlin's ubiquitous empti-ness", many artists, activists and other local residents attempted to *preserve* such empty spaces and, in the case of squatters, reanimate and rework them for alternative purposes (Weszkalynys, 2010: 62). The spatial practices of squatters were, in other words, part of a much wider struggle over the meaning and identity of Berlin in the years that immediately followed the fall of the Wall.

In more practical terms, the second major wave of squatting in Berlin drew on a sedimented history of contentious politics that encompassed both the modest acts of 'rebelliousness' undertaken by *Schwarzwohner* during the GDR and an action repertoire that emerged out of the squatting scene in West Berlin and else-where in West Germany and Europe during the 1970s and 1980s. If squatting thus served as a *necessary* protest *against* housing precarity, it also represented a *constituent* protest *for* alternative ways of living together. As in the case of their ear-lier counterparts, squatters who occupied houses in Berlin in the months following the collapse of the East German state were motivated by a desire to assemble a radical infrastructure that linked pressing local issues to wider articulations of autonomy and self-determination. A brochure produced in the summer of 1990 by squatters from a group of houses that formed part of the wider squatting net-work highlighted the range of practices and strategies adopted by their occupants. Echoing tactics first developed by Western activists in the 1970s, the introduction to the brochure describes a "growing movement" of young people who, in their own words, "want to realise their own ideas of collective living and working". "We want to break through the traditional structures of isolation," the authors added, "and build a self-determined neighbourhood infrastructure that prevents liveable homes from remaining empty and falling into disrepair (*eine selbstbestimmte Kiezkultur aufbauen und verhindern, dass bewohnbare Häuser leerstehen und verfallen*)."[39] At the same time, the new squatters were acutely aware of the rapidly changing circumstances in which they operated and that their actions should be seen as a challenge to the privatisation and financialisation of housing in the former East. As they concluded, "we cannot accept, without resistance, the actions of Western speculators who tear homes from our grasp only to turn them into marketable commodities".[40]

The various squatted houses profiled in the brochure not only signalled a further extension of existing occupation-based practices in Berlin, they also drew attention to the differentiation between squats in terms of their respective goals and motives (see Pruijt, 2013). For the eight activists occupying one wing of an apartment block at Prenzlauerallee 203/204, occupation represented far more than simply an opportunity for collective living. As they pointed out, "we started with the renovation of the house and repaired the roof. But we still have a lot in front of us to do [...] We want to take control of the house's energy needs according to sound ecological principles". In the case of another house at Rosenthaler Straße 68, it was three well-known bands from the underground scene in East Berlin – Ich-Funktion, Firma and Freytag – that were behind the occupation. According to the founders of what became known as the 'Eimer', "our wish is to build a permanent centre for art, culture and communications that enables a self-determined exchange with local and foreign groups while facilitating the development and exploration of various art forms." "Our own way of life," they concluded, "is to re-function the ruins of a vulnerable world and to create, in the process of living and working, our own culture."[41]

While squatted centres such as the Eimer, Kule ('Kultur und Leben'), and Tacheles undoubtedly played a decisive role in the cultural re-imagining of Berlin's empty spaces in the early 1990s, other squats documented in the brochure focused, on the development of local self-help initiatives and the assembling of familiar oppositional geographies. The 40 activists who squatted a house at Kastanienallee 85/86 described their own space as a "colourful and lively self-help project" with plans to expand the attic, open a print shop, a record store and other workshops. Similar plans were proposed by the squatters living in a house on Schliemannstraße. "We want to realise in our house," they proclaimed, "a collective form of life and work that preserves and extends existing neighbourhood structures. We intend to achieve this with the establishment of a children's club, a bicycle repair shop and a café."[42] Another group of squatted houses on Kreutzigerstraße in Friedrichshain drew up plans for a social centre that would include a cinema as well as a number of clubs and workshops while a house on

Lottumstraße was transformed into a radical meeting place for groups that had already been active in East Berlin in the final years of the GDR.[43]

The second major wave of squatting in Berlin was therefore dependent on the mobilisation of material and emotional geographies that built on a now well-established assemblage of spatial practices. The act of occupation was widely framed as a politics of adaptation, mending and rehabilitation as squatters tackled derelict spaces that (once again) required significant repairs. The squatters occupying a house on Tucholskystraße described, for example, how they entered a house "gray, dirty and dusty. The building structure is good. All the beams are solid holding everything in place. Step-by-step, we carried off all the dirt in order to make the house breathe again."[44] Improvised materials and do-it-yourself practices were, in this way, combined with grassroots political organising to form a makeshift urbanism that offered an alternative understanding of city life in a period of rapid social and economic change. Many squatters also saw their actions as offering a direct challenge to the privatisation of the housing market in the former East of the city as unification offered, for many, a new 'spatial fix' for the further urbanisation of capital and the creative destruction of socialist property regimes (see Heyden, 2008). At the same time, the tactics adopted by squatters were never crudely mimetic. While they drew on inherited forms of collective action, they also invented new ones. As scholars of contentious politics and social movements remind us, action repertoires are never wholly static even if they build on repertoires of action, claim-making and dissent that are well-rehearsed and conspicuously familiar (Tilly and Tarrow, 2007: 4, 11; Tarrow, 1998; Tilly, 2008). Repertoires, in other words, vary from place to place and time to time. For the squatters occupying empty houses in East Berlin in 1989 and 1990, it was the forging of new alliances and connections between activists from the West and the East that was central to the production of new spaces of solidarity whilst also producing a host of serious difficulties and problems.

The fall of the Wall, as recent scholarship has suggested, paradoxically prompted many Germans to feel deeply divided along East–West lines in spite of the fact that they were now citizens of a united Germany (see Berdahl, 1999; Borneman, 1992, 1993; Glaeser, 2000). "Unification of the nation," according to the anthropologist John Borneman, "involves suturing [...] two halves together, making real in the present the fantasy of a past unity, whether by the denial of difference, the annexation of one half by the other, or an incremental convergence of lifeworlds" (1993: 288). This was a process that often fostered a "profound disunity between East and West Germans". For many, the Berlin Wall that geographically and politically divided a country had simply been replaced by a new "wall in the heads of people" (Glaeser, 2000: 5). These divisions and differences were often experienced through the meaning and use of specific spaces. In the case of squatters, the occupation and construction of new geographies of protest and solidarity brought into sharp relief both the connections and tensions between activists from the East and the West. While the first squatted houses in late 1989 and early 1990 were predominantly occupied by groups from the GDR (especially in Prenzlauer Berg), the explosion of squatting in the spring of 1990 tilted the balance in favour of activists from the West. Some houses were made up of a mixed group of squatters though there were ultimately many fewer activists from the East.[45] As one former occupant of a house on Brunnenstraße remembered, "it was totally striking that there were only '*Wessis*' [in our squat]. We would have the odd token *Ossi* from time to time [...] At the time, I thought it was definitely noticeable that there were houses, like the Lottum [Lottumstraße 10a], where it was clear that everyone came from the East. And then there were houses like ours, where almost everyone came from the West. There was only a little mixing."[46]

If internal conflicts and disagreements are a frequent feature of social movements, it was ultimately the tensions between '*Wessis*' and '*Ossis*' that, according to many former activists, played a decisive role in the fracturing of solidarity during the second wave of squatting in Berlin (Gould, 2009: 329; see Feigenbaum et al., 2013; Owens, 2008). "I think the conflicts that forever existed between East Berliners and West Berliners arose out of the simple fact there used to be the Wall," wrote one former Mainzer Straße squatter. "This produced," he continued,

...very very different experiences. This was the first time that many from the East were able to openly defy what we commonly describe as 'structures.' Take, for example, the question of collective decision-making. Meetings were anarchic affairs. There were no structures and it was possible for everyone to have a say and decide for themselves. Any form of structure reeked, on the one hand of the Central Committee and power. On the other hand, many of the East Berliners showed a certain blind naiveté regarding the state and its intentions [...] They truly believed that the authorities meant well [...] This was unlike what the West Berliners brought in experience which had its own negative impact. This was often perceived as arrogant, know-it-all and a bit over the top (Arndt et al., 1992: 77).

The arrival of squatters from the West was eventually seen by many in the East as a form of 'invasion'. Many activists who had been involved in the political subcultures that emerged in the final years of the GDR drew parallels between the 'colonisation' of the former East and the subsumption and sidelining of one repertoire of contention by another. Differences invariably extended to the everyday practices of occupation. For a number of former squatters, it was an engagement with feminist politics which served as a source of confusion in squats that brought together activists from the East and the West (see Arndt et al., 1992: 91). Whilst heated discussions about squatting, activist spaces and the 'separation of the sexes' animated the autonomous Left in the West in the 1980s, they gained little traction in the East (see A.G. Grauwacke, 2003; Amantine, 2011). As one squatter recalled, "for us in the East, they were not an issue". According to another activist from the West, these differences were often overplayed and that it was "such a mind game anyway" while her comrades from the former GDR were far more "relaxed". "We were so totally uptight," she concluded.[47] And yet, for many women, the persistence of patriarchal attitudes and sexual violence within the German anti-authoritarian Left was also a source of serious conflict and friction. In 1990, a series of articles were published in the *BZ*, the main magazine of the new squatting movement, documenting the re-establishment of the notorious Indianerkommune in a house on Bergstraße in Berlin-Mitte. The Indianerkommune, which had been active in West Germany since the mid-1970s, openly promoted sexual relations between adults and children. In response, a group of autonomist feminists attempted to have the commune thrown out of the squat. In their view, the commune was a symptom of enduring patriarchal structures of domination and violence and that their fight against the sexual abuse of women and children was part of a wider movement of resistance.[48] In the eyes of many in the autonomist feminist scene, the *B-Rat* was woefully anaemic in its response. As they pointedly noted in another article published in the *BZ*, "it was clear that the *B-Rat* was full of types who could prattle on in assemblies about the need to tackle sexism. At the same time, they did not take any meaningful position or deal with the problem of sexual violence [...] Critical opinions only came from women (*Kritische Stellungsnahmen kamen nur vor Frauen*)."[49] Other queer activists tried to come to the defence of the commune arguing – rather misguidedly – that their work must be understood in terms of a much longer history of alternative youth projects and anti-authoritarian childrearing and that they hoped that a serious frank discussion about sexual politics was possible (Amantine, 2011; see also Herzog, 2005; Kommune 2, 1969).[50]

If the tensions between activists from the East and the West magnified existing fault lines within the activist community in Germany, feelings and practices of solidarity within the second wave of squatting were equally shaped by the micropolitical orderings of squatted spaces. Squats, as argued in Chapter 4, were intensely affective spaces that generated experiences of care and solidarity and moments of serious disagreement and dissent. According to one former squatter of a house at the corner of Neue Schönhauser Allee and Rosenthaler Straße, "the question of whether men and women could live together was never a big issue for us. Rather it was the remarkable situation that there were people in our house who called themselves squatters yet behaved in ways that contradicted this." Conflict often centred on the use of space. While many houses favoured the development of collective self-managed spaces, there were also houses that were occupied by relatively 'unpolitical' squatters who showed little interest in organising once legal contracts had been secured.[51] A number of squatters felt that the cultivation of modest geographies of connection and solidarity were easier when they were forced to negotiate internal and external struggles. "I think that we had less problems," one remarked, "when there was a lot going on around us – the whole political upheaval, reunification and getting to know the east." "And then," they added, "when it was quiet, so after three or four years, it kicked off internally." They were supported by another former activist who remembered how "after two years that things began to crumble with us. I think the reason was that our collective actually worked best while in an exceptional state of crisis". Such a view was further reinforced by a fellow comrade who noted how "from the very beginning we were already thinking about how to live collectively with a communal kitchen, shared spaces, etc. This wasn't an intellectual exercise. We just wanted to cook together (*Man wollte halt zusammen kochen*). And when the pressure from the outside decreased, when, in other words, we all had contracts, there was a pronounced change in our collective life".[52]

In this way, the material and emotional life of squatted houses in the former East was shaped by changing internal dynamics. At the same time, the connection to wider neighbourhood geographies and histories was far more limited and febrile than it had been during the first major wave of occupation in Kreuzberg in the 1970s and 1980s. While there were indeed many former *Schwarzwohner* who were part of the underground network of protest and dissent that emerged during the final years of the GDR, there were also many others who saw their actions as individual acts of autonomy and rebellion rather than practices that were embedded in a wider repertoire of contention. Others still framed their actions in terms of a need to secure affordable housing for themselves and showed little interest in the articulation of wider movement goals while, for the growing number of squatters from the West, the Eastern half of Berlin represented wholly unfamiliar territory (Holm and Kuhn, 2011). Even in the spring of 1990, the outline of the Wall remained and West Germans were required to show their passports when crossing into the East. As one squatter recalled, "in May, it was still the GDR here". "In the beginning," he continued, "what was cool, as a Westerner in the Eastern part of the city for me and for others, was a total rethinking of who we were" (quoted in Arndt et al., 1992: 74; see Gutmair, 2013).

Unsurprisingly, perhaps, there were significant tensions between squatters (especially those who had come from the West) and other local residents in East Berlin. The power vacuum that accompanied the fall of the Wall precipitated a further rise in far-right violence across East Germany and in Berlin, in particular, where neo-Nazi activists had occupied a number of houses on Weitlingstraße in the district of Berlin-Lichtenberg (Brand and Fregonese, 2013). Squatted spaces were subject to repeated attacks in the early months of 1990 by fascist youth as inner-city districts became sites of violent contestation over who, in fact, was able to stake a claim to the city. Many squatted houses were forced to become defensible spaces.[53] Windows were barred and doors barricaded. In the words of one former squatter, "we simply want to build our own habitus. In our community, a predominantly leftist potential thrives. For right-wing groups and fascists, we have naturally become a target. We want to live here and need to ensure that we safely can" (Arndt et al., 1992: 96). While squatters concentrated on the defence of existing houses they also joined with other militant anti-fascist organisations from both the West and the East and were eventually successful in forcing the group of neo-Nazis to abandon their base on Weitlingstraße.

For many locals, however, it was the activities of squatters that were responsible for attracting far-right groups and that it would be better for the local
neighbourhood if the squatters left. Squatters – most notably those who arrived
from the West – were often seen as 'colonisers' who simply jumped the queue
for housing in neighbourhoods where some residents had waited for up to 10
years to receive their housing allocation. As one activist later remarked, "the
'Easterners' [*Ostler*] in our area felt especially cheated."[54] To the wider
community, the arrival of squatters from the West came as a shock. According
to one squatter, "in the Mainzer Straße and the surrounding area, there were
undoubtedly people who initially did not know what would come over from the
West. They certainly hoped for something completely different, perhaps even a
Mercedes. And as we were the first, they obviously got a fright. I can understand that too. I can understand that people were afraid of us with our green
hair" (quoted in Arndt et al., 1992: 85).

Efforts were made by a number of squatted houses to foster better relations
with the local neighbourhood though attempts to establish a broad infrastructure
of alternative practices and services were largely unsuccessful. Unlike Kreuzberg
in the 1970s and 1980s, squatted spaces in the former East remained relatively
isolated and were unable to scale-up their activities and produce new geographies
of engagement and solidarity. These challenges were felt by squatters from both
the West and the East. Whilst some houses actively promoted relations with their
neighbours, organising open houses, street festivals and concerts, others looked
inwards and turned their attention to the development of alternative forms of
collective living. "The result," as one squatter concluded, "was that you may have
tried here and there to socialize. On the whole, however, it was a case of living
side by side with an abundance of misunderstandings." "We were living," another
activist added, "[...] in a parallel society [*Parallelgesellschaft*] par excellence."[55]
The disconnection between squatters in East Berlin and the wider community
was a view that was also shared by many neighbourhood residents. For one
member of a local citizen's initiative that had sprung up in response to the occupation of houses on Mainzer Sraße in April 1990, the squatters were less interested in the articulation of a right to a different city than in the establishment of
a militant political movement. "It would be a lie," she noted, "if we now said that
we welcomed that [the arrival of squatters]. They did not move in to live here, but
rather to quickly build their own scene." "We had no radar for each other," she
continued. "Neither us for the young squatters [...] or indeed them for us. They
were not willing to accept what we wanted [...]." Another member of the same
initiative was similarly antipathetic. "We naturally wondered," she recalled, "how
can you live there without having to seek a lease agreement, without otherwise
paying for energy, gas, etc. They wanted to created communal spaces and tore
down walls to do so. One can do that but if I tear down a wall, I would re-plaster
everything in order to make it look like a home" (quoted in Arndt et al., 1992:
142, 143, 156).

To be sure, the squatting scene that sprang up in East Berlin after the fall of the Wall was not viewed only with singular. It also received significant support from the local community and the numerous underground movements that had emerged during the final years of the GDR. It was clear, however, that there were significant differences between the second major wave of squatting in East Berlin and its predecessor in the West. As Andrej Holm and Armin Kuhn have argued, "the squats of the 1980s were part of an extended and differentiated alternative subculture that centred on the inner-city districts of Kreuzberg and Schöneberg" (2011: 652). The squatting movement during the period was embedded within a wider ecology of protest and resistance that had its origins in the anti-authoritarian revolt of the late 1960s and early 1970s. Squatters were, in other words, able to articulate a critical urbanism that depended on a vast archive of practices that promoted a different way of inhabiting the city. The squats in the 1990s, in contrast, were unable to establish themselves as a coherent, broad-based movement and were largely confined to isolated pockets of dissent in a city that was undergoing rapid change.

And yet, failure was not inevitable. Whilst scholarship on social movements has increasingly acknowledged the importance of geography to contentious politics, it has often located decline within a common, even fatalistic, cycle of protest. Such a view occludes the geographical complexities behind the formation and dissolution of activist movements (see Tarrow, 1989, 1998; Voss, 1996). As this chapter has argued, the second wave of squatting drew on a historical geography of protest whose origins lie on *both* sides of the Wall. The squatters who occupied houses in the former East during the *Wende* adopted tactics developed by *Schwarzwohner* during the GDR as well as an action repertoire that played a decisive role in an earlier wave of occupation in West Berlin. While the new scene experienced its own series of internal disagreements and disputes, it was also able to articulate an understanding of shared city life that occupied (quite literally) the cultural, social and political vacuum that was a defining feature of Berlin in the months and years after the collapse of the GDR. Ultimately, it is in this context that the eviction of the Mainzer Straße squatters and the eventual fracturing of the scene must be understood. The police operation that expelled the squatters was, if nothing else, characterised by its violence as the squatters were removed and their spaces demolished and ransacked. In one of the most striking and poignant scenes in Juliet Bashore's 1991 documentary, *The Battle of Tuntenhaus*, the American filmmaker returns to Mainzer Straße in the immediate aftermath of the eviction only to find West Berlin police officers looting a former squat and queer social centre (Bashore, 1991).

The sociologist Saskia Sassen (2014) has recently argued that "our advanced political economies have created a world where complexity too often tends to produce elementary brutalities" (2). There is much to recommend in this view though we may wish to ask where it is, in fact, violence and brutality that ultimately serve as the necessary precondition for the further urbanisation of capital? The squatting movement in the former Eastern half of Berlin may have been unable, in the end, to construct a broad alternative infrastructure that connected activists to other forms of grassroots political organising. It was also, however, *forcibly expelled* from many houses as the city of Berlin underwent a period of intense urban restructuring within which the activities of squatters were seen as a major obstacle to be both pacified and proscribed. And yet, while the second wave of squatting in Berlin may have ended with the Mainzer Straße clearances in November 1990, the dissolution of the scene as a coherent 'movement' was also accompanied by the emergence of new *experimental geographies* in Berlin that adapted and reworked the tactics and strategies of urban squatting as a means of reclaiming a renewed right to the city (Lefebvre, 1996).

Postscript

In August 1990, a group of 20 squatters entered the canteen of the *tageszeitung* (*taz*) newspaper on Kochstraße in broad daylight and dismantled a table that had originally belonged to Kommune I in the late 1960s. A couple of days later, members of the Berlin press received an invitation to a squat on Mainzer Straße where they were presented with the table in its new location. While the *taz* impugned the squatters for their actions describing them as "brats" [*Rotzlöffel*], the squatters argued that they were simply freeing the table from its increasingly corporate setting.[56] The table was itself an important material artefact of the New Left in West Germany. It had been purchased in the spring of 1969 as part of the establishment of the Socialist Lawyers' Collective (*Sozialistische Anwaltskollektiv*) by Horst Mahler, Klaus Eschen, Hans-Christian Ströbele und Ulrich K. Preuß. It soon found its way to the home of Kommune I on Stephanstraße in Berlin-Moabit – the infamous commune that counted Dieter Kunzelmann, Rainer Langhans and Fritz Teufel amongst its members. In 1974, the commune had become the '*Sozialistische Zentrum*' and the table the centre, so to speak, of intense political debates about the reorganisation of the Left in West Germany. *Rote Hilfe*, *Schwarze Hilfe* and the precursors to a number of other political movements including the German Green Party all used the table for their meetings. With the founding of the *taz* in 1978, the table was donated to the editors as the newspaper was widely seen as the "heir to the student movement". As Hans-Christian Ströbele noted in a letter to the *taz* in September 1990, the table was never the sole property of the *taz* but rather had been entrusted to them.[57]

In many respects, the confiscation of the table by the Mainzer Straße squatters was simply the latest stunt within a repertoire of political performances that had played a central role in the long development of the anti-authoritarian Left in West Germany and in West Berlin in particular. For the squatters, it served as an assertion of their own position as the latest representative of the much wider history of radical politics explored in this book and for which the city of Berlin had become a key theatre of protest. It also marked, in their view, a break with that history and a new beginning which was symbolised by the table's 'reclamation' and 'restitution' within the squatting scene in the former East. The table's new home was short-lived, however, and as the eviction of the squatters on Mainzer Straße threatened, the table was dismantled again and shared between different houses before it was secretly moved. It later ended up in an occupied villa on the outskirts of Berlin before it was stolen by squatters from Potsdam in 1993 who finally set it on fire.[58]

It would, of course, be tempting to treat the Kommune I table and its social life as a microcosm of the shifting historical dynamics of the New Left in Germany. But what perhaps matters more here is the centrality of *geography* to the creation and circulation of "new activist imaginations and the development of new ways of configuring political relations and spaces" (Featherstone, 2012: 6). Just as the table was assembled and dismantled, reassembled and moved, so too was the case for an extra-parliamentary opposition that depended on the continuous production of new geographies of engagement and solidarity that were firmly rooted in an understanding of urban space as a key site of action and revolt. As I have argued in this book, this is an understanding of protest and resistance that prompted Berlin's squatters to construct spaces that re-imagined the city as a "flexible resource" for other forms of political, social and economic organisation (Simone, 2008: 200). While the actions of squatters were ultimately unsuccessful in sustaining a city-wide movement for alternative forms of housing and dwelling, their fortunes, it seems to me, were rather different from that of an object which in many ways came to anatomise an entire geography of protest. Unlike the table, which was eventually destroyed, the attempt to re-imagine the city of Berlin as a crucible of political change did not come to an end with the violent clearing of squatters. As the next chapter shows, the tactics and strategies of urban squatting were adapted and reworked by groups of activists and artists as well as a host of other organisations in the city. The spatial practices mobilised by squatters thus transformed Berlin into a living 'archive' of alternative knowledges, materials and resources. It is this archive that continues to play a central role in the struggle for a more radical and socially just urbanism.

Notes

1   The term '*Schwarzwohnen*' offers up no easy equivalent in English. '*Schwarz*' is the German word for 'black' and the well-known term '*schwarzfahren*' refers to the practice of fare dodging on public transport. The term '*stilles besetzen*' (covert squatting) was also used in the GDR. See Mitchell, n.d.: 6.

2   Robert-Havemann Gesellschaft (hereafter RHG), "Behörden erneut gegen instand-besetzte Lychener 61", *Umweltblätter*, Nr. 4 (1988), p. 2.

3   'Samizdat' is the Russian word for self-publishing and came to be used by dissident circles in the GDR (*Samisdat*) to describe the production and distribution of subversive texts.

4   As the Umwelt-Bibliothek was officially part of the Zionskirche parish, it was legally protected by the state-church compact which covered the church's internal publications. Samisdat publications that were produced under the auspices of various parishes were marked for "internal church use only" ("*Innerkirche Information*"). See Glaeser (2010: 449–450).

5   RHG, "Lychener Straße 61 in Berlin-Prenzelberg am Ende?" *Umweltblätter*, Nr. 10 (1988), pp. 2–3; "Durchsuchung", *Grenzfall*, Nr. 14 (1987), p. 2. *Umweltblätter* was replaced by the journal *Telegraph* which ran articles in 1995 on the history of squatting in the East Berlin.

6   Landesarchiv Berlin (hereafter LAB) C Rep. 100-05 Nr. 1472.

7   While these numbers remain an estimate, there is considerable evidence, as historians such as Udo Grashoff and Peter Mitchell have argued, to support them. Mitchell points, for example, to the fact that there were 646 recorded cases in East Berlin in the first six months of 1984; in the same period for the subsequent year some 483 apartments had been illegally occupied. Grashoff, in turn, shows that, by the beginning of 1985, the tenure status of over 1270 apartments in Prenzlauer Berg was unclear (Grashoff, 2011b: 19; Mitchell, n.d.: 4). In reviewing these numbers at the Landesarchiv in Berlin, it is clear that there are some discrepancies between various districts and the city more generally. See LAB C Rep. 111 Nr. 57, "Stellvertreter des Oberbürgermeisters für Wohnungspolitik: Information über die Eingaben während der Vorbereitung der Wahlen am 6.5.84", "Ungesetzliche Bezüge im I. Halbjahr 1984"; LAB C Rep. 100-05 Nr. 1996, Magistrat von Berlin, Büro des Magistrats, "Stand der Erfüllung der Wohnraumvergabepläne der Räte der Stadtbezirke per 21. Aug. 1985", 2.10.85; LAB C Rep. 134-02-02, Nr. 1358, Rat des Stadtbezirkes Prenzlauer Berg, Ratsitzung am 21.11.1985, "Ungeklärte Mieteingänge per 1.1.1985", n.p.

8   Interview with J.L. (July 2013).

9   Interview with J.L. (July 2013).

10 RHG, *Umweltblätter*, Nr. 7 (1987), p. 2f.

11 Papiertiger Archiv (PTA) "Hausbesetzer Ost", *Instand-Besetzer-Post*, 19.6.1981, p. 6.

12 Once again official records provide some clues. There were only 88 forced evictions carried out in East Berlin in 1983, despite the fact that 954 cases of illegal squatting were recorded in the city over the course of the year. See LAB C Rep 100-05 Nr. 1945/1, "Eingabeanalyse 1983", pp. 7–8. A more detailed picture of forced evictions in East Berlin is, however, still needed.

13 An article in the *Instand-Besetzer-Post* talks of fines of 300 M. See PTA, "Hausbesetzer Ost", *Instand-Besetzer-Post*, 19.6.1981, p. 6.

14 It is clear when looking at the records of the various district councils in East Berlin that housing insecurity and evictions were often a product of changing personal circumstances (death of a relative, divorce, illness and institutionalisation). See, for example, LAB C Rep. 135-02-02, Nr. 1179, Rat des Stadtbezirks Friedrichshain, Ratsitzung am 23. November 1978, "Räumung einer Wohnung gemäß § 23 der Verordnung über die Lenkung des Wohnraums Betrifft"; LAB C Rep. 135-02-02, Nr. 1186, Rat des Stadtbezirks Friedrichshain, Ratsitzung am 1. Feb. 1979, "Zustimmung zur Räumung von Wohnungen auf dem Verwaltungswege gemäß §23 Verordnung Wohnraumlenkung (12/79)."

15 PTA, "Hausbesetzer Ost", *Instand-Besetzer-Post*, 19.6.1981, p. 6.

16 PTA, "Hausbesetzer Ost", *Instand-Besetzer-Post*, 19.6.1981, p. 6.

17 RHG, "Lychener Straße 61", *Umweltblätter*, Nr. 10 (1988), p. 2f.
18 There were large squatting scenes in other East German cities most notably Leipzig and Potsdam.
19 PTA, *Interim*, 11.1.1990; Hamburger Institut für Sozialforschung (hereafter HIS) "Über die Hausbesetzerbewegung in Ost-Berlin (hereafter HBB), Teil 1", *Telegraph* 9 (1995), pp. 38–40; see also "Gesprächsrunde über die Ostberliner HausbesetzerInnenbewegung in den 1990er Jahren," http://www.squatter.w3brigade.de/content/gespraeche/da-haben-die-die-ganze-h%C3%BCtte-besetzt-0 (hereafter SQ1).
20 It is now clear that there were a number of far-right *Schwarzwohner* in Berlin-Lichtenberg occupying houses on Weitlingstraße before they were officially 'squatted' in 1990.
21 HIS Archiv, "HBB, Teil 2", *Telegraph* 10 (1995), pp. 36–37.
22 In the first few months after the fall of the Wall, it was still illegal for West Germans and other 'Westerners' to stay overnight in the East. Nevertheless, many did so by crawling through holes in the former Wall (see Gutmair, 2013: 46).
23 HIS Archiv, "HBB, Teil 1", *Telegraph* 9 (1995), p. 40.
24 These plans were spearheaded by companies that had played a role in the marketisation of squatted spaces in West Berlin in the 1980s.
25 HIS, "HBB, Teil 3", *Telegraph* 11/12 (1995), p. 36.
26 PTA, *Interim*, 26.4.1990.
27 HIS, "HBB, Teil 3", *Telegraph* 11/12 (1995), p. 37.
28 HIS, "HBB, Teil 3", *Telegraph* 11/12 (1995), p. 37.
29 Interview with C.L. (August 2009).
30 PTA, letter dated 26.6.1990 from Vertragsgremium to Berlin Magistrat, Ordner Häuserkämpfe O-Berlin.
31 HIS Archiv, "HBB, Teil 1", *Telegraph* 9 (1995), p. 43.
32 PTA, report dated 9.7.1990 for B-Rat, Ordner Häuserkämpfe O-Berlin.
33 HIS Archiv, "HBB, Teil 1", *Telegraph* 9 (1995), pp. 43–44.
34 RHG, Boxfile Runde Tische RTc-01.
35 HIS, "HBB, Teil 1", *Telegraph* 9 (1995), p. 45.
36 HIS, "HBB, Teil 3", *Telegraph* 11/12 (1995), p. 41.
37 HIS, *Telegraph* 11/12 (1995), pp. 41–33.
38 PTA, *BZ* Nr. 14, 28.11.1990, pp. 4–5.
39 PTA, brochure, *Hausbesetzer, Selbstdarstellung von 16 Projekten*, Ordner Häuserkämpfe O-Berlin.
40 PTA, ibid., Ordner Häuserkämpfe O-Berlin.
41 PTA, ibid., Ordner Häuserkämpfe O-Berlin.
42 PTA, ibid., Ordner Häuserkämpfe O-Berlin.
43 See http://bandito.blogsport.de/uber-uns/
44 PTA, brochure, *Hausbesetzer*, Ordner Häuserkämpfe O-Berlin.
45 Interview with C.L. (August 2009).
46 Quoted in SQ1, "Gesprächsrunde über die Ostberliner HausbesetzerInnenbewegung in den 1990er Jahren."
47 Ibid.
48 PTA, *BZ*, 10.10.1990, pp. 12–14.
49 PTA, *BZ*, 5, n.d. 1990, pp. 15–16.
50 PTA, flyer, "Gegenmacht", Ordner Häuserkampf Berlin-Mitte.

51 SQ1, "Gesprächsrunde über die Ostberliner HausbesetzerInnenbewegung in den 1990er Jahren."

52 Ibid.

53 See HIS, "HBB, Teil 2", *Telegraph* 10 (1995).

54 SQ1, "Gesprächsrunde über die Ostberliner HausbesetzerInnenbewegung in den 1990er Jahren."

55 Ibid.

56 HIS Archiv, "HBB, Teil 3", *Telegraph* 11/12 (1995), pp. 39–40.

57 *taz*, 3.9.1990.

58 HIS Archiv, "HBB, Teil 3", *Telegraph* 11/12 (1995), pp. 40–41.

Chapter Six
Capture and Experimentation

They say gentrify, we say occupy.
Graffiti in Berlin-Neukölln, Schillerpromenade (2013)

Metropolitan revolt is always a refoundation of the city.
Antonio Negri (2002)

This chapter begins with the story of another eviction. On 2 February 2011, an alternative housing project in the Berlin district of Friedrichshain, Liebig 14, was cleared in a police operation that involved several thousand officers.[1] Liebig 14 was one of the last surviving projects within Berlin's rapidly disappearing autonomist scene that included a handful of houses on Liebigstraße and the neighbouring Rigaer Straße (Boyle, 2011). These houses formed part of an alternative milieu that combined a commitment to autonomous self-management with a radical anti-capitalist politics. They also played an important role in the emergence of a broad grassroots movement in Berlin that encompassed a range of city-wide issues that included increasing rents, the displacement of low income residents, and the gentrification of inner-city neighbourhoods (Bernt and Holm, 2009; Colomb, 2012; Holm, 2010; Holm, 2014b; Holm and Kuhn, 2011; Novy and Colomb, 2013; Scharenberg and Bader, 2009).

The planned eviction began in the early hours of the morning of 2 February as police units arrived on the scene only to be welcomed by hundreds of protestors.

Metropolitan Preoccupations: The Spatial Politics of Squatting in Berlin, First Edition.
Alexander Vasudevan.
© 2015 John Wiley & Sons, Ltd. Published 2015 by John Wiley & Sons, Ltd.

The number of supporters soon rose to over 1000 and, as the police struggled to control the protestors, the order was given to clear the building. Specially-trained officers entered the building at 08:00 through the ground floor and the roof. They soon encountered a series of makeshift barricades, overflowing bathtubs and a missing stairwell that had been destroyed by the occupants. In the end, it took the police over four hours to successfully evict the nine remaining residents (see Boyle, 2011).[2] At the same time, units outside the house battled an ever-growing crowd of supporters. Luxury apartments in the surrounding neighbourhood were vandalised, as were a number of storefronts. As night fell, a solidarity march through Friedrichshain ended in further clashes between protestors and the police. Banks, supermarkets, department stores and the O2 Sports Centre were all targeted. In the end, over 80 arrests were made, while a preliminary police report estimated that the damages would exceed 1,000,000 Euros (Boyle, 2011).[3]

In many respects, the Liebig 14 eviction represented the culmination of a long history of protest and resistance, occupation and experimentation. The house had originally been occupied by squatters in the early months of 1990 as part of a major wave of squatting that accompanied the fall of the Wall and the collapse of the GDR. In the aftermath of the violent eviction of squatters on Mainzer Straße in November 1990, the occupants of Liebig 14 entered into negotiations with the owners of the house, the *Wohnungsbaugenossenschaft Friedrichshain* (or WBF), and the City of Berlin. It was only in November 1992 that the squat was finally legalised as an alternative housing project and a total of nine units in the house were able to sign lease agreements which, in practical terms, guaranteed the house's status as a self-organised collective (see Lennert in SqEK, 2014: 81). Renovations began on the house which were finally completed in 1996. In 1999, the house and a number of neighbouring projects were purchased with a view to their further 'modernisation' as an eco-residential block for young professionals. Negotiations between the new owners and the residents of Liebig 14 broke off in March 2001, and over the course of the next 10 years a protracted battle over the status of the house and its occupants ensued which finally led to the early morning eviction in February 2011.[4]

The story behind the eviction of Liebig 14 brings together a number of themes that are at the heart of this chapter; namely, the transformation and re-orientation of Berlin's squatter movement around a network of legalised spaces and projects; the 'capture' and instrumentalisation of occupation-based practices by the state; the relationship of squatting to gentrification and other forms of urban restructuring; and the further creative destruction of Berlin's housing stock and the concomitant emergence of a new radical housing movement in the city. The chapter shows how the tactics and strategies of urban squatting were adapted and reworked by groups of activists and artists as well as a host of other organisations in Berlin from the early 1990s to the present. It argues that the long and complex history of squatting transformed the city of Berlin into an active archive of alternative knowledges, materials and resources which remained central to the development of new forms of collective living. At the heart of this chapter, therefore, is a concern with the enduring significance of 'occupation' as a political process that re-casts the city as a site of radical social transformation (see Vasudevan, 2014a). In other words, the chapter explores the possibilities of an alternative ontology of the city and the relationship of squatting to a range of different occupation-based practices that speak to both basic rights claims and demands (housing and infrastructure) and prefigure other ways of thinking about and inhabiting the city (autonomy and self-management).

In order to do so, the chapter examines a broad repertoire of practices that illustrate the importance of reclaiming a renewed right to a different city. More specifically, it develops three interrelated perspectives that emerged directly from the squatter movement in Berlin in the wake of the evictions on Mainzer Straße. First, it discusses the role of cultural experimentation and artistic practice in the development of new strategies for participatory architecture, community design and everyday dwelling (see Heyden, 2008; Vasudevan, 2011a). Second, it explores the establishment of an "unusually stable" ecology of protest that provided an opportunity for the articulation of new political horizons and the cultivation of other identities and intimacies (Lennert in SqEK, 2014: 81; see also Amantine, 2011, 2012; Azozomox, 2014a, 2014b). Third, it considers the recent emergence of a new round of housing activism in Berlin that has come to challenge gentrification, dispossession and rising rents (new forms of citizen 'occupation', protest camps and eviction resistance). Taken together, the chapter concludes by showing how the practice of urban squatting continues to offer conceptual tools and practical resources through which a more radical and socially just urbanism may be produced.

There is admittedly a danger here in romanticising the figure of occupation and the spatial grammar that it presupposes. After all, and as this book has already shown, the squatting movement in Berlin in both its major incarnations ultimately failed as a broad-based struggle for an alternative urbanism. And yet, the normative demand for a different vision of the city – one no longer shaped by the disagreeable materialities of displacement and dispossession – remains, for many Berliners, a pressing issue. As this chapter argues, the return to 'occupation' was never a crude repetition of an earlier repertoire of contention, especially in an era where the expulsion of people from their homes and wider lifeworlds has become an increasingly central feature of contemporary urban life (Delclós, 2013; Novak, 2014; Sassen, 2014; Schling, 2013). At stake here, is a further re-imagining of the act of occupation "as a political process that *materialises the social order which it seeks to enact*" (Vasudevan, 2014a: 6; emphasis added). Reclaiming a right to housing thus represents far more than a "re-politicisation of urban politics" as Andrej Holm has recently suggested (2014b: 23). It also carries wider geographical implications. As this chapter shows, the common work of occupation involves not only different ways of extending bodies, objects and practices into space but a set of orientations that insist on the very possibility of a radically different understanding of what it means to think about and inhabit cities.

Squatting as Art: Performing Architectural Activism in the New Berlin

For many in Berlin's squatter movement, the violent eviction of squatters on Mainzer Straße represented a turning point and a key moment in the movement's decline. While solidarities between squatters, occupied houses and local neighbourhoods had already begun to fray and unravel in the immediate lead-up to the events of November 1990, the brutality of the expulsions was greeted by an overwhelming sense of despair, exhaustion and powerlessness (see Arndt et al., 1992; Bashore, 1992). As this book has argued, squatted spaces were sites of collective world-making and their dissolution was, unsurprisingly, characterised by a shared sense of loss. And yet, for squatters and other activist communities, the sense of loss that accompanied the fracturing of solidarities was never wholly debilitating or de-politicising (Gould, 2009: 396). For Judith Butler, loss – as a felt experience – was equally *constitutive* of new social, political and aesthetic relations however fragile and precarious (2002: 467). Butler acknowledges that loss, on the one hand, represents a form of *un-making*. "There is," she writes, "the loss of place and a loss of time, a loss that cannot be recovered or recuperated (2002: 468). On the other hand, if "places are lost – destroyed, vacated, barred", their disappearance becomes, according to Butler, the very condition for the articulation of new political horizons. Butler's argument chimes in this way with recent scholarship on forced evictions as "embodied, located and grounded phenomenon". As the geographer Katherine Brickell has recently argued, the emotionally saturated experience of eviction often animates rather than inhibits the production of new geographies of activism and engagement (Brickell, forthcoming; see also Delclós 2013; Novak, 2014; Romanos, 2014).

In the case of the squatting scene in Berlin, the months that followed the clearance on Mainzer Straße were shaped by a number of dynamics. Many who were evicted from houses on Mainzer Straße left the wider movement – some permanently, others for a shorter period of time – as they struggled to cope with their own feelings of anger and despondency. A number of activists also found refuge in the remaining squatted spaces that persisted in the wake of the evictions. At the same time, there was also a growing recognition that further violent confrontation with the police would only lead to another round of expulsions. As city-wide structures including the *Besetzerrat* (B-Rat) and the *Vertragsgremium* splintered, new attempts at a negotiated settlement with municipal authorities were undertaken. These efforts represented, if anything, a 'scaling down' that sought out new geographies of connections in local neighbourhoods and drew on the experiences of squatters who were predominantly from the former GDR.[5] In Prenzlauer Berg, a number of houses had already responded to the increasing ineffectiveness of the B-Rat by forming a new 'squatting round table' (*Runden Tisch Instandbesetzung*) that brought squatters together with municipal authorities, district politicians and the *Wohnungsbaugesellschaft*, the local neighbourhood successor to the KWV.[6] The round table met, in fact, for the first time on 9 November 1990 just a few days before the police operation on Mainzer Straße. Little progress was, however, made during the first meeting as a number of political parties refused to attend. The following meetings that took place in the aftermath of the Mainzer Straße evictions brought a new urgency to the negotiations. As the squatters faced the prospects of changing legislation and a tide of property claims from former owners, negotiations were seen as a pragmatic measure that would secure the future for squatted houses while providing spatial continuity for a wider network of oppositional activities.[7]

After two months, a settlement was agreed in principle in January 1991 that came close to the original position of the *Vertragsgremium* though, in practice, squatted houses were obliged to sign lease agreements as housing projects under which individual tenancies were legally embedded.[8] In this way, some squatters were also able to secure tenancies which guaranteed that they were paying rents at cheaper East German rates.[9] Whilst the squatters in Prenzlauer Berg were roundly impugned by their colleagues in other districts across the former East for making common cause with a corrupt system, it soon became clear that their modest success offered an opportunity to avoid further criminalisation and eviction. As a group of squatters in Mitte noted, "a Berlin-wide approach to negotiations with the Magistrat [...] simply did not work." "[*The Vertragsgremium*]," they concluded, "was no longer capable of articulating a realistic perspective especially after the events of the past few months."[10] By the end of January 1991, a new round table for squatters in the district of Mitte ('*Runden Tisch Instandbesetzung Mitte*') was therefore established, while a project group for squatted houses in Friedrichshain ('*Projektgruppe besetzte Häuser Friedrichshain*') was formed in the middle of March.[11] Negotiations, however, once again stalled, prompting many houses across Berlin to secure individual agreements at the expense of the wider 'scene' and it was the informality of these ad hoc legal arrangements that would later play an important role in the recent wave of evictions that targeted house projects such as Liebig 14 (Azozomox, 2014a; Holm, 2014b: 14). Houses unwilling to negotiate were cleared, the last wave of evictions taking place between 1996 and 1998 under the direction of former Bundeswehr general, CDU hardliner and local politician, Senator Jörg Schönbohm.[12]

The legalisation of squats in the former Eastern half of Berlin thus contributed to the formation of a new and uneven geography of protest in the city. Houses soon doubled as social centres within a stable yet fragile infrastructure of alternative practices that subsisted in neighbourhoods undergoing rapid change. While many squatted spaces were quickly subsumed within the wider logics of urban regeneration, there were also other sites that became crucibles of intense cultural experimentation. For a number of students studying at the Hochschule der Künste in Berlin, for example, it seemed clear that new practices and tactics were needed in the wake of the Mainzer Straße evictions. To them, claims for a "transformed and renewed right to urban life" (Lefebvre, 1996: 158) could no longer depend on entrenched forms of militancy. What was needed, they believed was a new form of

Figure 6.1   Performing architecture? Squatters occupying Kastanienallee 77, Berlin-Prenzlauer Berg on 20 June 1992 (pamphlet produced by K77).

activism that attempted to balance and *extend* claims for self-determination with greater cooperation with local authorities (see Vasudevan, 2011a).

Recalling the links between activism and the performing arts described in Chapter 3, the group of students at the Hochschule adopted a new form of site-specific practice which, as one former squatter noted, served as a "catalyst for the occupation of abandoned buildings". On 16 December 1990, they occupied an empty apartment in Friedrichshain and turned its spaces into a temporary gallery space housing the First Mainzer Art Exhibition. A second exhibition was held on 25 February 1991 in various rooms of a gallery in Kreuzberg. Further 'happenings' and occupations ensued until 20 June 1992 when a number of activists dressed as doctors and nurses occupied one of the oldest buildings in Prenzlauer Berg at number 77 Kastanienallee (Figure 6.1).[13] The building had been empty for six years. Originally built in 1848, the 3-storey building predated the Hobrecht city plan and sat on an unusual 10m × 100m lot. The 'complex' consisted of three houses separated by three interior courtyards. To squat here, as the group would later proclaim, was to respond to a "medical emergency" and save "the heart of the house, dress and heal its wounds, and fill it with life".[14]

The actions of the squatters returned, in this way, to an earlier mode of anti-authoritarian performance (Boyle, 2012; Kraus, 2007: see especially Chapter 3).

Drawing explicit inspiration from the work of the German artist Joseph Beuys, the group which took over Kastanienallee 77 (hereafter K77) in 1992 deliberately recast the act of squatting as a form of "continuous performance (*unbefristeten Kunstaktion*)" or installation art. K77 became, in the words of Beuys, a *social sculpture*, a location for "non-speculative, self-defined, communal life, work, and culture" (Heyden, 2008: 35).[15] According to Beuys (quoted in Tisdall, 1974: 48), his objects were to be understood as "stimulants for the transformation of the idea of sculpture [...] or of art in general. They should provoke thoughts about what sculpture can be and how the concept of sculpting can be extended to the [...] materials used by everyone". To think of sculpture as constitutively *social* was to therefore draw attention to the different practices through which "we mould and shape the world in which we live".

Beuys's working methods became something of a credo or manifesto for the group of activists that had come to work and live in K77 and it is perhaps not surprising that, over the course of the summer of 1992, a number of varied performances, exhibitions, and installations were created. "We made theatre (*wir haben Theater gemacht*)" were the words of one former occupant, while another described the occupation as a "theatrepiece (*Theaterstück*)" that built on recent developments in performance art.[16] As he pointed out, "there was no plan (*es gibt kein Plan*)" or set of rules governing the squat. "Every space could be played with," added another founding member of the house. "The possibilities were endless." For many, these were possibilities that transformed the building into a "*Freiraum*" or "free space" that demanded creative experimentation.[17] It was only with late summer rain and colder weather that the realities of living in a building that did not have a roof, proper windows, or water, gas and electricity set in (Figure 6.2).

Experimentalism quickly shaded into pragmatism. Without any financial or legal support, producing a 'social sculpture' depended upon the constructive use of 'found materials' as well as improvisatory 'improvements' to the building's existing form. At the same time, to secure more permanent residency, the group adopted a similar approach to other squatted spaces in Prenzlauer Berg. They initiated negotiations with local authorities, working hard to acquire legal status which they attained in 1994. A 50-year lease was signed and a communal, "non-property oriented solution to ownership" was resolved through the creation of a foundation which channelled profits into a number of socio-political projects, both in Berlin and the developing world (Heyden, 2008: 35). The foundation running K77 was also successful in securing public funds via the Structural Self-Help initiative, though this only covered 80 percent of the reconstruction costs. As a former inhabitant recalled, "the remainder was made up through our own contribution. We all toiled up to 50 hours a month over three long years on the building site."[18] The building was, in this way, painstakingly renovated. The state of the house meant that the entire building was gutted. Almost every ceiling, wall and window was replaced while a completely new system of pipes for heating, water and electricity was

Figure 6.2   The 'art' of squatting (pamphlet produced by K77).

installed. Sustainable planning principles were used, recycled building materials adopted and strict conservation laws closely followed (Heyden, unpublished MS).[19]

Over 100 people have lived in K77 over the years since its initial reoccupation. Today, approximately 25 adults and children still live together "in one flat" across six levels in three different buildings. Seventy percent of the complex is now devoted to living arrangements (Figure 6.3). The other 30 percent is set aside for the house's various cultural projects and includes a non-profit cinema, a ceramics workshop, studio space, and a homeopathic clinic. The core of the project remains the negotiation and transgression of boundaries (political, social, cultural) and the creation of what Mathias Heyden, one of the original occupants of K77, has described as an "architecture of self and co-determination [that] questions the right to the design and use of space" (unpublished MS). For Heyden, K77 remains something of an architectural laboratory for user participation and self-organisation. Heyden's own description of the project indeed highlights the role

Figure 6.3   Kastanienallee 77 in 2009 (photo by author).

of the built form in creating new "dwelling perspectives" (Ingold, 2000) that are themselves dependent upon the unpredictable evolution of spaces. According to Heyden (2008: 35–36),

> Every two years, the inhabitants [of K77] sort out who wants to live where and in which constellation, so that the usage and interpretation of available spaces is constantly renewed [...] In the process-oriented planning and building stage, a broad variety of forms of participation and self-organization came about: the new spaces were largely laid-out through flexible and self-built wallboards. Wall partitions were accordingly fitted with omissions. Light openings, room connections, or breaks in the wall were designed so that they can be closed and reopened at any time. Overall, design decisions were left to individuals [...]

A number of former occupants singled out in turn the kitchen as the key "socio-spatial centre of the house". The same floor also contains a communal dining room, a room for children to play and a "bathing landscape". More general issues relating to the *ongoing* and collective redesign of K77 from floor layouts to infrastructural 'improvements' are discussed and agreed on by all members of the house. Such attempts to foster a sense of collective property and economy coincide with a strong commitment to overcoming "particular conditionings of the individual and the self" (Heyden, 2008: 36, 37). To do so is to work towards the construction of a new habitus, to borrow Bourdieu's now overworked term (see Bourdieu, 1977). In the particular case of K77, it is the very *performance of architecture* itself that has become, in this context, a key source of inspiration for a whole host of self-organised and collective everyday practices that conjoined infrastructure and community (Heyden, 2008).

K77 can therefore be seen as the spatial manifestation of "a much broader understanding of self-empowered space" (Heyden, 2008: 37). It was also, for many years, part of an informal network of squatted houses, all located in former districts of East Berlin, to which the development of *shared* cultural spaces was a common cause. The network included K77 and houses at Auguststraße 10 ('KuLe'), Kleine Hamburger Straße 5, Lychener Straße 60, and Rosenthaler Straße 68 ('Eimer'). At stake here was an understanding of 'occupation' as a radical cultural practice, a form of 'architectural activism' that offered a critical point of purchase on new strategies for participatory architecture, community design and spatial appropriation (Heyden, 2008: 37). For Mathias Heyden, the "emancipative social sculpture" of K77 – and for that matter the tactics and practices of squatting more generally – represented only one example or possibility of how an embodied and practical understanding of the built environment was crucial to the design of potential spaces for future commons (Heyden, 2008: 37). Indeed, there remain a number of projects, groups and networks in Berlin that have attempted to explore these issues across a range of sites from sculpture parks and temporary event-based installations to more systematic attempts at community design (Fezer and Heyden, 2004; Haydn and Temel, 2006; Till, 2011). And yet, the 'use' of temporary spaces, abandoned buildings and disused lots for the development of alternative practices and events has also played a growing role in Berlin's ongoing neo-liberal restructuring. As Claire Colomb and others have persuasively argued, "from the early 2000s onward, [...] the creative, unplanned, multifaceted, and dynamic diversity of such 'temporary uses of space' was gradually harnessed into urban development policies and city marketing campaigns" (Colomb, 2012: 132; Novy and Colomb, 2013; Cupers and Miessen, 2002).

While planners and policymakers in the immediate post-*Wende* era saw Berlin's countless empty spaces as the ruinous remainders of unwanted pasts and contemporary economic failures, many others – including squatters, artists and marginalised social groups –saw these "indeterminate spaces" as laboratories of autonomy, experimentation and self-determination (Groth and Corijn, 2005: 503). The development and emergence of an alternative network of underground urban activities was, in other words, closely intertwined with the use of such interstitial in-between spaces. This took on a number of different forms that included squats, caravan communities, community gardens, flea markets, raves, techno clubs, art installations and waterfront beaches. What quickly became known in German as *Zwischennutzung* ('interim or temporary use') encompassed a wide range of activities which took place in spaces whose use was "planned from the outset to be impermanent" and where a sense of 'temporariness' was actively promoted (Colomb, 2012; 135; Haydn and Temel, 2006: 17). These were, as the geographer Karen Till has argued, "interim spaces" that were widely seen as sites of escape and refuge, transgression and possibility (2011: 106).

In the years immediately after the fall of the Wall, such interim spaces in Berlin were predominantly run by actors that came from outside "the official, institutionalized domain of urban planning and urban politics" (Groth and Corijn, 2005: 506). Over the course of the last decade, however, the temporary use of vacant buildings and spaces has increasingly captured the attention of politicians, planners and developers. This has led, in particular, to the establishment of a policy regime in the early 2000s based on the explicit mobilisation and integration of temporary uses and interim spaces into official planning initiatives and wider place-marketing discourses (Colomb, 2012: 138). The shift in local economic development policies was characterised by a new focus on 'cultural revitalisation' and 'creativity-led' policies as a key strategy for enhancing the city's global brand image. Becoming a 'creative city' was increasingly seen by many in Berlin as the necessary catalyst for future economic success and it was in this context that local politicians sought to capture and harness alternative subcultural spaces as a competitive advantage set against a backdrop of intensifying interurban competition (Mayer, 2013b: 4). Temporary uses and interim spaces were therefore marketed as sites that served as potential playgrounds for creative workers and firms directly or indirectly involved in the culture industries, and in many cases they also operated as tourist attractions in their own right (Colomb, 2012: 138). In a more practical register, the city developed a series of formal strategies to encourage the formation of new creative clusters, while the temporary use of different spaces was facilitated by a relaxing of licensing and planning procedures. A 'Temporary Use Agency' (*Zwischennutzungagentur*) was set up in order to encourage the use of empty retail spaces in deprived neighbourhoods across Berlin. At the same time, the operating terms of the *Liegenschaftsfonds*, the state-owned entity that was responsible for the sale of public property, were relaxed to allow "temporary use contracts" for "non-profit, community-oriented activities" on sites that were in public ownership (Colomb, 2012: 139, 140).

It is perhaps hardly surprising that many local officials viewed temporary use arrangements as simply the best option in the absence of any other development opportunities and that they offered, if anything, a stepping stone for future profit-making initiatives. As Claire Colomb has rightly argued, it is telling that the city's own study of temporary spaces was titled 'Space Pioneers' (*Raumpioniere*) and drew attention to "a new species of urban players, for whom urban spaces, untamed territory at best, is something to be discovered, squatted, conquered" (Misselwitz et al., 2007: 104; Colomb, 2012: 141). If the terms used by city planners invoked a rather different understanding of 'occupation', they were also, according to Colomb, startlingly redolent of the language used by scholars such as Neil Smith in order to describe the process of gentrification (1996). Temporary uses and interim spaces were, in this way, valued above all else as a 'means to an end' rather than for any intrinsic use value. Such practices were, on the one hand, vital for the production of spaces of social, cultural and artistic experimentation. On the other hand, they also acted as a catalyst for more conventional forms of urban redevelopment that threatened the very survival of these spaces, forcing them to adapt, relocate or disappear. As Jamie Peck (2012a) has suggested, the wider discourse around the creative city represents a "largely symbolic but nevertheless consequential" set of practices that have provided local officials with the language and rationale to frame and enforce pre-existing neoliberal policies (464).

While many squats did not see themselves as temporary spaces in any strict sense, a focus on artistic activities and other cultural practices was singled out by local officials – as in the case of the Tacheles building – as a fitting example of a young, vibrant and creative city. Such a relationship between radical political spaces and local planning initiatives has prompted Colomb to argue that Berlin policy-makers were forced to negotiate a "delicate balancing act" between the 'use' of alternative social movements and the cultivation of mainstream urban development strategies (2012: 143). This may in some sense be true but is it not also the case that radical spaces such as squats found themselves in precarious positions as both islands of alterity and resistance *and* agents of gentrification and urban regeneration? Were squatters still able, in other words, to articulate an alternative right to the city? Recent scholarship on neo-liberalism and the post-political city has pointed, in this context, to a decline in the possibilities for "transgressive politics in European and North American cities" (Van Schipstal and Nicholls, 2014: 174). The violent predations of capital combined with the further consolidation of private property regimes and the emergence of neo-liberal forms of urban citizenship have shown that the very "right to have rights" in the city is now firmly tethered to a demonstration of economic value (Arendt, 1973; see Van Houdt et al., 2011). According to this view, squatters and other radical urban groups are framed as either a threat to the wellbeing of the larger community or a source of informal urban practices that, taken together, serve as a blueprint for the further commodification of urban space (Balaban, 2011; see also Holm and Kuhn, 2011; Iveson, 2013; Mayer, 2013a; Uitermark, 2004, 2011).

In the end, over the past two decades, squats, legalised house projects and other social centres have continued to survive in central districts of Berlin in the face of commodification and criminalisation, co-optation and stigmatisation (Van Schipstal and Nicholls, 2014: 175; see SqEK, 2013, 2014). Whilst the strategic options available to squatters have undeniably shrunk and continue to shrink, the recent history of occupation-based practices in Berlin demonstrates an enduring desire to defend and sustain existing spaces and solidarities. At the same time, it also highlights a growing willingness amongst many activists to seek out new openings and connections and recognise the constraints and possibilities that shape and condition their activities. At stake here is a critical urbanism that is able to interrogate and challenge dominant conceptualisations of the city – especially as they have come to coalesce around notions of 'creativity', 'integration', and 'liveability' – and to show that alternative imaginings are not only possible but also desirable (see Uitermark, 2011).

From Creativity to Experimentation

As the preceding pages have shown, the long history of squatting in Berlin has continuously produced and sustained new orientations toward urban living. While this a history that encompasses a rich and complex set of practices, the opportunities to occupy abandoned buildings and empty spaces and carve out a radical urban politics have become increasingly limited over the past 20 years. Squatters have responded by pursuing a number of different strategies that have tended to operate in the space *between* informality and legality, creativity and resistance (see Van Schipstal and Nicholls, 2014; Holm, 2014b). In practical terms, this has been characterised by the extension of a longstanding repertoire of contention that sees the act of squatting as the 'other' to creative destruction. The legalisation of squatted houses in the early 1990s thus represented a pragmatic compromise for activists faced with the prospect of imminent eviction. It also allowed for the establishment of a network of radical spaces across the city of Berlin that continued to challenge capitalist economic and social relations (Lennert in SqEK, 2014: 81).

And yet, after the evictions on Mainzer Straße and the legalisation of a number of squats, it also became clear that formerly squatted spaces were often detached and isolated from a new regime of urban renewal. If some squats offered a patina of cultural glamour and edginess, their special status as affordable housing projects also fomented neighbourhood conflicts over displacement, privatisation and rising rents that made cooperation with tenants and citizen initiatives more difficult (Holm and Kuhn, 2011: 654; see also Holm, 2006; Rada, 1991). It would be misleading, therefore, to speak of a broad-based squatter movement in Berlin in the 1990s and 2000s. The city's redevelopment regime had, if nothing else, succeeded in disarticulating the spatial practices of squatters from other urban protest movements and networks. Squatters thus faced something of a double bind.[20] Whilst the continued existence of alternative housing projects represented an opportunity to experiment with radical forms of shared living and working, it also reminded squatters that their survival stemmed, in no small part, from the cultural capital they conferred on an increasingly neo-liberal city.

It is, however, possible to detect a number of recent developments within the Berlin squatting scene that have come to challenge a narrative that saw a movement in terminal decline and at the end of its life cycle. In contrast, and echoing work on the history of squatting in Amsterdam (Owens, 2008; see also Pruijt, 2003; Uitermark, 2004), squatters in Berlin were able to reconfigure the political and cultural dimensions of their practices in terms that allowed for the production of new geographies of action and engagement (Staggenborg, 2001). Over the past couple of decades, this has taken on a number of forms that can be plotted along at least two axes. First, the relative stability afforded by legalisation provided a context that transformed collective spaces into emancipatory sites where new identities and intimacies were actively produced. Squats and their legal offshoots often became spaces that *queered* the home as a traditional site of domesticity and social reproduction (Amantine, 2011, 2012; Azozomox, 2014a). Second, the political content of the wider squatting scene shifted to other causes, movements and networks that demanded new forms of practice and solidarity. The politics of the 1990s differed significantly from the 1970s and 1980s and the resources and infrastructures developed by squatters soon assumed a new organising role that supported the development of the alter-globalisation movement as well as other transversal alliances and connections (Owens, 2008: 227). In this way, squatters in Berlin were able to find new ways to survive in a wider urban environment that was increasingly antipathetic to their activities. While squatting as a political movement in the city was therefore constituted as a fundamental critique to the urbanisation of capital, it also represented an opportunity to develop autonomous, feminist, anti-fascist, queer and other subversive forms of politics. These struggles continued to shape the composition of the scene in the 1990s and they remain an important factor in the political trajectories of various house projects across Berlin (Amantine, 2011; Azozomox, 2014a).[21]

The development, for example, of an autonomous feminist politics in West Germany during the 1970s provided inspiration for a number of women-only initiatives such as those discussed in Chapter 4. The second wave of squatting after the fall of the Wall initiated a similar number of new projects. Some groups were evicted after a short period of time including houses at Mariannenstraße 9–10, Dieffenbachstraße 33 and the women-only house on Mainzer Straße (no. 3) that was cleared as part of the police operation in November 1990. Other spaces have, however, survived including a feminist-lesbian-trans house at Brunnenstraße 7, a queer-anarcho-feminist house project at Liebigstraße 34 and a women-only space at Grünberger Straße 73 (Azozomox, 2014b: 191). These spaces provided a radical platform to explore and extend a number of practices and strategies that first emerged in the late 1970s as a way of making visible new forms of resistance against patriarchal power structures, endemic sexual violence and other shared forms of oppression (Amantine, 2011: 125).[22] This encompassed struggles against everyday sexism, particularly in the private sphere, and an active critique of heteronormative systems in house projects and other social centres. It also involved an ongoing questioning of established bipolar sexual identities and assignments as well as a commitment to autonomous self-determined forms of housing (Amantine, 2011: 67–68).[23]

In the final instance, it depended on a sustained engagement and working through of a wide body of feminist theory and practice that brought together concepts of gender performance and intersectionality with radical queer and autonomist positions (see Amantine, 2011; Dennert et al., 2007; Schultze and Gross, 1997).

One of the key reasons for the development of autonomous feminist spaces was a desire to challenge "male-dominated behaviour and patterns in mixed living spaces, squats or political groups, where often men ignored and rejected the need for independent women's spaces" (Azozomox, 2014b: 192). It was in this context that Liebig 34 was first squatted in 1990 as part of the wider wave of occupations that took place during the *Wende*. The squat was legalised and it is now home to 35 residents from across the world who describe the project as a "queer-anarcho-feminist collective". For the occupants, the use of the word 'queer' is deliberate. In their own words, it represents a "denaturalisation of normative concepts of masculinity and femininity, the decoupling of gender from sexuality, the destabilisation of the binaries of hetero -and homosexuality, as well as the recognition of a sexual pluralism that also includes bisexuality, transsexuality, transgender, inter-sexual and gender queer [thus] broadening the limitations of gay and lesbian fixed identification" (quoted in Amantine, 2011: 105). As a queer space, the residents of Liebig 34 have self-consciously attempted to forge a prefigurative geography that anticipates a collective form-of-life whilst disrupting the traditional family structure. As a 'separatist space', the house is exclusively occupied by feminists. There are no men although the project is involved in activist work with other mixed groups. As the residents make clear, separatism does not involve the projection of traditional 'male' roles onto women but rather the "abolishment of a sex specific division of labour". In their own words, a "separatist life free of sexist norms is created in which feminists of every age can open up, develop and support each other in the process of emancipation. Through the form of workshops and working groups, we always try to strengthen and diversify the skills and talents of all of us."[24]

Despite the efforts of house projects such as Liebig 34 to articulate a radical queer politics that valorises an autonomist feminist position, violence against women has, as the writer and activist Amantine argued, remained a constant feature of the squatting movement in Berlin from the fall of the Wall to the present (see Amantine, 2011). In 1990, for example, a serious conflict broke out in a squatted house on Marchstraße in Berlin-Charlottenburg between three men and many of the women living in the house who accused the men of unwanted sexual advances and sexist behaviour. At one point, the wing of the house occupied by the women was attacked and bombarded with bricks, stones and Molotov cocktails. The three men were eventually forced to leave.[25] Similar violent confrontations took place in other houses as well. In August 1992, a house at Kinzigstraße 9 in Berlin-Friedrichshain was the scene of violent street battles between punks and squatters who occupied different parts of the house.[26] Six years later, a serious conflict arose between squatters in a house at Brunnenstraße 6–7 in Berlin-Mitte as four men were kicked out of the house for sexual harassment. The house was fire-bombed in retaliation by a group that included some of the men and their supporters while the fallout within the wider autonomist movement prompted many women to once again challenge the "laddish posing and endemic power structures" that were part of the house and the scene more generally.[27] The same year, a woman was sexually assaulted at the Tommy Weisbecker-Haus, a former squat and social centre in Kreuzberg. As an open letter to the house by a group of women made clear, the reaction of the house's occupants "fluctuated between silent toleration and open support for the attacker". In 2000, another open letter was sent to the Köpi, a former squat and housing project in Berlin-Mitte, in response to an assault on a woman by one of the residents, highlighting the need for clear anti-patriarchal guidelines and that such incidents should be acknowledged and made public (Amantine, 2011: 144).[28]

The development of projects and spaces such as Liebig 34 should therefore be seen as a response to the persistence of uneven power structures and systemic forms of violence within the wider squatting scene in Berlin. There were, unsurprisingly, similar challenges for other queer spaces within the movement. The second 'Queen House' (*Tuntenhaus*) which was cleared as part of the Mainzer Straße evictions was soon reformed in the wing of a squat on Kastanienallee in Berlin-Prenzlauer Berg and is now a legal house project. Unlike its predecessors, the project is less overtly political although it remains connected to other initiatives including the Gay-Antifa (*Schwule-Antifa*) and the newspaper *Tuntentinte* (Azozomox, 2014b: 199). The queer caravan site, *Schwarzer Kanal*, which was first squatted in 1989 and is part of a network of trailer encampments (*Wagenburgen* or *Wagenplätze*) across the city, also continues to play an important role in the squatter movement (see Van Schipstal and Nicholls, 2014).[29] The encampment has been forced to move twice and is now located in the district of Treptow. The site remains a key part of Berlin's autonomist/queer/trans scene and is responsible for organising and hosting events such as the Queer and Rebel days as well as an annual DIY Radical Queer Film Festival (Azozomox, 2014b: 199).[30]

In the end, a commitment to the development of radical political spaces that acknowledge other identities and intimacies is now widely accepted within the former squatting movement in Berlin as well as other ecologies of protest and resistance across the city. This is not to gainsay the fact that the dissolution of sexism, racism, trans- and homophobia has not always been reflected in practice. The challenges that former squats and house projects face in the struggle to become autonomous safe spaces reflect the complex process of collective identification through which new radical selves and solidarities were openly, passionately and often precariously enacted. Reports in 2009 of the persistence of "patriarchal and hierarchal structures, sexist speech and behaviour" as well as serious sexual and physical violence within a squat at Brunnenstraße 183 in Berlin-Mitte only served to reinforce the fears of many activists within the local activist community. As a statement released by *Wir Bleiben Alle* (WBA), an umbrella organisation formed in the 1990s for squatters in Berlin, highlighted, "despite a turnover in occupants, attacks have continued with little response from other residents. It therefore seems all the more important to us that the structures and processes which led to these incidents are investigated by the house and the WBA and that those affected are protected and left alone." "The WBA," the statement continued, "apologizes [...] that nothing was done to intervene even though the situation in B183 [Brunnenstraße 183] was well-known. It is clear to us that we have not even succeeded in leftist projects to create shelters and spaces of care for victims of abuse." The WBA recommended the formation of an 'awareness group' as the starting point for a debate on sexism and hierarchies within leftist networks and structures and a "small step on the road to free open spaces" (einen kleinen Schritt auf dem Weg zu Freiräumen).[31]

The ongoing struggle for the right to assemble and sustain such spaces in Berlin was therefore shaped by tensions that have played a central role within the evolution of the anti-authoritarian Left in Germany (see A.G. Grauwacke, 2003; Amantine, 2011). But it was the question of geography – its ever-evolving significance for the development of radical spaces of care, connection and engagement – that represented the key driving force behind the various forms of political and cultural action within the squatters' movement in Berlin. Whilst the place-based nature of the movement was characterised by local struggles over the right to housing and infrastructure and a desire to forge other forms of shared city life, by the late 1990s the focus of the movement had also shifted to new modes of action. Cultural experimentation took on, in this way, a new significance as the city's commitment to the culture industries created the necessary conditions for institutionalisation and, in some cases, co-optation. At the same time, the movement adopted other forms of protest that reflected wider political trends. As illegal squats gave way to legal house projects, it would be tempting to detect a de-politicisation of the movement. What had changed, however, was less a question of movement decline or depoliticisation sensu stricto but rather an awareness that the political terrain had moved onto new causes and issues (see Owens, 2008: 227). In particular, the squatting movement was increasingly tied to anti-globalisation struggles as well as new initiatives against labour precarity and the privatisation of social services (López, 2013). These developments transformed squatted spaces, social centres and house projects into sites that provided the resources conducive to the building of new practices and solidarities across Berlin.

It is in this context that the geographer and activist, Anja Kanngieser (2012), has zoomed in on the concept of 'transversality' as a way of accounting for the emergence of new subjectivities, relations and worlds within and beyond existing forms of political action. As Kanngieser shows, the concept of transversality was first developed by the French philosopher, psychiatrist and activist, Félix Guattari, as a way of exploring the production of radical collective subjectivities. For Guattari, transversality represented a tool for re-imaging power structures in institutions and was a product of his work at the experimental La Borde psychiatric clinic in France (see Dosse, 2011). According to Guattari, it was possible to overcome dominant hierarchical structures by exploring different forms of connection between groups and their environments and that, within particular institutional settings, other organisational innovations could be assembled that were non-representational, non-programmatic and *spatially generative*. Kanngieser argues that such an understanding of transversality has important consequences for re-imagining the textures of radical political organisation. In practical terms, she examines the tactics of collective and visible re-appropriation that have been gaining attention over the past decade, most notably from a network of activists connected to autonomous and precarious movements across Europe and Latin America (Kanngieser, 2007, 2013; see also Raunig, 2007, 2010). At the heart of these movements is a preoccupation with the subversion of capitalist logics of exchange in favour of a concept of appropriation predicated on need and desire and unhindered by economic constraints. This included, for example, the *Umsonst* ('For Free') campaigns that took place between 2003 and 2006 in Berlin and elsewhere in Germany and which consisted predominantly of gatherings composed of undefined activists and a wider public collectively and playfully engaging in a range of illegal acts (fare evasion, trespassing, theft) in the social realm (Kanngieser, 2007).[32]

As Kanngieser shows, these campaigns placed particular emphasis on the development of creative gestures of resistance and liberation that connected direct action with a wider commitment to public engagement. This was characterised, according to Kanngieser, by a shift in tactics. In this way, new forms of cooperation and solidarity were composed that adopted a relatively open and flexible approach to organisation as well as new inventive techniques that "instigated movements across differential social groupings and structures, art and politics, urban spaces, and political nodes and institutions" (Kanngieser, 2012: 269). By creating the necessary conditions for these encounters and events, groups such as *Umsonst* were able to produce what Brian Reynolds has elsewhere described as a "transversal territory", a "catalyzing and transitional space from which new experiences, subjective reconfigurations, and, by extension, dissident mobilizations can emerge" (2009: 287). For the remnants of Berlin's squatting scene, these actions also represented the possibility for the generation of new common spaces and public realms that moved beyond traditional movement goals and strategies. As the rapid transformation of Berlin in the 1990s increasingly foreclosed the opportunities for squatting, the cultivation of transversal links between former squatters and other social groups was crucial in sustaining spaces and sites of self-managed activity. What is pivotal here is a recognition of the value of constructing "new relationships and connective junctures between people and environments" that agitate against the logics of capitalist accumulation in its various forms. In the case of many formerly squatted spaces and house projects, these new initiatives connected houses to other "interventions in precariousness, privatisation and gentrification" that, taken together, formed the basis of a revivified alternative urbanism across the city of Berlin (Kanngieser, 2012: 284). If the traditional status of the squatting scene within neighbourhoods had been weakened, such geographies of connection, movement and processuality ultimately laid the groundwork for the emergence of a new round of housing activism in Berlin.

Reclaiming the Radical City

The recent history of Berlin is a story indelibly marked by the ongoing neo-liberal transformation of the city. In a book documenting the dynamics of urban regeneration in East Berlin, the urbanist Andrej Holm offers a forensic dissection of the changing social geography of the district of Prenzlauer Berg in the decade or so that followed the fall of the Berlin Wall (2006). As Holm argues, urban regeneration, large-scale modernisation, and the advent of a new property regime precipitated the widespread displacement of existing residents who were no longer able to afford rising rents. By 2001, over 40 percent of tenants living in flats that had been 'modernised' had moved out of the district (Holm, 2006: 220; see Rada, 1997). They were largely replaced by young urban professionals for whom Prenzlauer Berg increasingly offered the social and cultural 'benefits' of inner-city living. Existing state support for urban regeneration was, in turn, increasingly financialised as direct state subsidies earmarked for repairs and upgrading were replaced by indirect incentives and wider market imperatives (Holm, 2006: 206–208).

While Prenzlauer Berg represented something of a laboratory for gentrification in a newly unified Berlin, it was by no means unique. As Holm recently suggested, the city's official housing strategy over the past of couple of decades amounted to nothing short of a blueprint for the development of a *neo-liberal city*. Fiscal instability in the early- to mid-1990s served, in this way, as a pretext for the privatisation of Berlin's social housing stock and the concomitant abandonment of social housing subsidies which were withdrawn in 1998 for new social housing construction. Follow-up subsidies for housing built between 1987 and 1997 were terminated in 2003. During the same period, the city government promoted the sale of its own housing stock. Over 200,000 social housing units have since been privatised, and between 1998 and 2004 two state-owned housing companies with over 40,000 and 65,000 units respectively were sold to private investors including large private-equity firms (see Fields and Uffer, forthcoming; Uffer, 2014). At the same time, building regulations were liberalised just as the ability of local authorities to plan and build new homes was drastically curtailed in favour of private developers (Holm, 2014b: 6, 8). It is perhaps not surprising, therefore, that rents across the city have risen by over 25 percent since 2003 with inner-city neighbourhoods recording even higher increases.[33] A recent survey showed that, on average, just 27 percent of all family-sized apartments or houses across Germany could be covered by rents that represented a third of the average household net income. For low-income households (60 percent of the local average), the number of affordable homes dropped to 12 percent (Holm, 2014a: 17). The effects of these developments have only served to exacerbate existing social inequalities and spatial divisions as the 400,000 or so households in Berlin under the official poverty line faced further pressure to meet basic needs. The impact of rent hikes also coincided with changes to state benefits – most notably the Hartz IV reforms – which, in practical terms, meant that full rents were no longer covered by benefits because income thresholds, according to new regulations, exceeded the cost of accommodation.[34] Many tenants soon fell into arrears which was used by owners as a pretext for terminating lease agreements and issuing eviction notices. Between 2009 and 2012, the Berlin Courts recorded on average 6000 evictions a year although housing experts have argued that an equally large number of evictions have gone unreported.[35] Whatever the case, over 5 percent of all house moves in Berlin in 2011 were a result of evictions, forced or otherwise (Holm, 2014b: 10, 11; see Novak, 2014).

It is against a backdrop of dispossession and displacement that the baleful 'redevelopment' of Berlin must therefore be understood. As Holm persuasively argues, "gentrification is not a special case of urban regeneration, but rather the *new urban mainstream*" (2014b: 11; emphasis added). And yet, the further neoliberalisation of Berlin has not gone uncontested and, if anything, the last five years have borne witness to the emergence of new broad-based oppositional movements in the city against rising rents and forced evictions, in particular, and gentrification more generally. While citizens' initiatives sprang up in a number of neighbourhoods that have a long history of activism including Kreuzberg, Neukölln and Schöneberg, new tenant groups also formed in other neighbourhoods including Pankow ('*Pankower Mietenprotest*'), Alt-Treptow ('*Kiezinitiative Karla Pappel*') and Moabit ('*Runder Tisch gegen Gentrifizierung*'). In many respects, these initiatives reactivated an older tradition of urban activism embedded within local neighbourhoods. At the same time, however, they also connected up with recently established forms of protest against gentrification, precarious work and the privatisation of the city's utilities (Holm, 2014b: 12; see Scharenberg and Bader, 2009; Novy and Colomb, 2013). As Margit Mayer (2013a) has pointed out, these developments have come to represent a new and tentative bridging of traditional activist practices and groupings (which included squatters) with the victims of a neo-liberalising urbanism characterised by austerity measures, the deregulation of labour markets and new repressive strategies from stricter laws to tougher policing (12).

In many respects, it was the plans for a mega-development scheme in the recently amalgamated district of Kreuzberg-Friedrichshain that served as a catalyst for new forms of protest in Berlin. The Media Spree development focused on the transformation of both banks of the river Spree into a creative cluster that traded on a dense local web of artistic, musical and subcultural activities (Novy and Colomb, 2013: 1825). By 2008, opposition to the project had precipitated the formation of a loose network of activist groups under the banner "*Mediaspree Versenken*" ("Sink the Media Spree"). The coalition of activists agitated against the project's master plan which they believed would lead to the privatisation of access to the river, the gentrification of local neighbourhoods and the co-optation of various alternative subcultures. As Albert Scharenberg and Ingo Bader (2009) have argued, what distinguished the protest was its ability to bring together a range of different actors from a nucleus of hardened activists and squatters based in Kreuzberg to beach bar owners and other entrepreneurs, from neighbourhood tenant organisations to groups representing the city's underground clubbing scene (332).

The movement adopted a creative repertoire of contention that combined neighbourhood 'walks' and community workshops with artistic installations and an aggressive social media campaign. At the same time, activists collected signatures over a six-month period which triggered a non-binding public referendum in the district of Kreuzberg-Friedrichshain.[36] The referendum was held on 13 July 2008 and, while it was framed in terms of specific planning criteria, activists used it as a means of derailing the wider Media Spree project. The results exceeded the campaigners' expectations with a turnout of 19.1 percent and a recorded approval rate of 87 percent. As the most successful referendum in the history of Berlin, local municipal authorities were forced to back-pedal and examine alternative options.[37] Negotiations were initiated with activists though many aspects of the plan were declared "non-negotiable". This led to internal disagreements within the wider movement as solidarities fractured between different groups over the effectiveness of the negotiations (Novy and Colomb, 2013: 1825, 1826).

The protests have nevertheless continued as part of the 'Mega Spree' initiative which represents a similarly broad coalition of actors and groups "affected by current processes of restructuring and gentrification" (Mega Spree, 2010). If the initial protests were widely trumpeted as a success, however modest, they did so by demonstrating that, in Berlin, the articulation of an alternative right to the city remained a fragile possibility. If anything, recent struggles have gravitated from a preoccupation with resisting the capture and enclosure of insurgent creative practices towards new mobilisations against housing and labour precarity and against the discrimination and disenfranchisement faced by many marginal groups in the city (Mayer, 2013a: 12). At stake here is a form of resistance to an urban landscape where "capital as a force of creative destruction" has become a palpable presence and where, for many inhabitants, the balance between creativity and destruction has ineluctably tilted toward the latter (Harvey, 2014: 78; see Peck, 2012a). In the case of Berlin's formerly squatted spaces, many of which had been legalised in the 1990s, the development of new alternative urbanisms only served to reinforce an increasingly paradoxical predicament. On the one hand, the consolidation of a neo-liberal policy regime in Berlin meant that many house projects were no longer tolerated and were under threat of imminent eviction. On the other hand, the re-emergence of a radical housing politics in the city was characterised by a return to occupation-based practices as part of a new spatial grammar of dissent and resistance.

Between 2009 and 2012, a number of prominent house projects in Berlin were cleared by the police, including Brunnenstraße 183, Liebig 14 and the Kunsthaus Tacheles. In each instance, the eviction was a product of the complex legal wranglings that led to the normalisation of squatted houses in the early 1990s but was equally responsible for loopholes that undermined those very agreements (Holm, 2014b: 14; see Azozomox, 2014a). This was further complicated by the changing ownership of the houses which fluctuated during a period where Berlin's housing market was fully financialised and exposed to a growing global crisis. It is against this very backdrop that the original legal basis for many of the houses was challenged as tenants and projects were issued with eviction notices and expelled from their homes. The forced evictions of Brunnenstraße 183 and Liebig 14 in 2009 and 2011 respectively each involved a large police operation, while the Kunsthaus Tacheles was peacefully cleared in September 2012 after years of legal and economic uncertainty (Azozomox, 2014a: 276).[38] Liebig 14 has since been 'modernised' by the owners. Brunnenstraße 183 is currently under redevelopment and the Kunsthaus Tacheles remains an empty architectonic symbol of displacement, profiteering and speculation (Figure 6.4). A number of other houses and projects remain under threat, including former squats at

Figure 6.4   The 'remains' of the Brunnenstraße 183 squat after its eviction in 2009 (photo by author).

Brunnenstraße 6/7, Reichenberger Straße 63 A and Linienstraße 206. The former DDR dissident organisation Kirche von Unten also faces eviction from its home in Prenzlauer Berg and even houses that were legalised in the early 1980s as part of the first wave of squatting in Berlin have been threatened with rent hikes and possible evictions (Azozomox, 2014a: 295, 296).[39]

Andrej Holm is therefore right to sound a sombre note regarding the possibility for new rounds of urban squatting in Berlin. As he points out, the occupation of the south wing of the former Bethanien Hospital in Kreuzberg in June 2005 and the subsequent formation of the New Yorck social centre remains the only successful occupation in the city over the past decade (2014a).[40] The occupation emerged in response to the forced eviction of a legal house project at Yorkstraße 59 on 6 June 2005.[41] The eviction once again revealed the violent bullying tactics of (new) owners who resorted to often illegal methods of 'mobbing' – defined here as a form of psychological and emotional bullying – to threaten and traumatise the residents of the project (see Molé, 2012; see Rothe, 2013). This included turning off the water to the house, disconnecting phone lines, walling up doorways, stealing mail, installing surveillance cameras and repeatedly abusing and threatening the residents. Three years after the eviction, the Berlin Court of Appeals ruled that the eviction was illegal and that all charges that had been laid against the occupants were to be rescinded (Azozomox, 2014a: 279, 280).

Unlike earlier attempts at squatting, the occupation at the Bethanien complex was not immediately cleared by the police as local politicians expressed a willingness to negotiate with the squatters. The Bethanien complex already enjoyed a long history within the wider activist scene in West Berlin as the site of the first squat in 1971, the Georg von Rauch-Haus. District politicians were therefore reluctant to create a political flashpoint though negotiations nonetheless broke down and it was only new municipal elections in 2006 that brought both groups back together as part of a round table initiative to resolve the impasse over the occupation. Four years later, a 15-year lease was finally agreed under a new trusteeship model and the immediate future for the New Yorck as a neighbourhood social centre was secured (Azozomox, 2014a: 280, 281). If the house represented a singular success for squatters and other activists, it did not represent the only effort to connect a right to a different city with the production of a "critical geography of occupation" (see Vasudevan, 2014a). In fact, a number of recent examples have demonstrated that the logic of occupation continues to resonate across an expanded field of dissent and resistance, autonomy and self-determination.

In 2012, for example, a community centre and recreation home that had been used for over 15 years by old-age pensioners in the Berlin district of Pankow was occupied by a large group after the local council said that it would have to make way for a new luxury real-estate development.[42] The squatters, aged between 63 and 96, were part of a community of over 300 pensioners who paid a monthly membership of 1 Euro in order to use the space for a range of activities (Amantine, 2012). The space had become a site of care, gathering and refuge for many older

Germans from the former East who found themselves unemployed after the *Wende*. Whilst the occupation began as a local protest against the district council, it quickly mutated into a broader struggle against the vagaries and predations of late capitalism. As one of the occupants recalled in an interview, "the most important thing is to stay together – here or elsewhere. We have the impression that you all think we are a bit senile. We are not young but we are not the grannies from 30 years ago sitting on a chair by the stove and knitting socks for their grandchildren."[43] "We do not want to leave them [our children and grandchildren] a country," added the squatters in a joint press release, "in which a music lesson for children, a visit to the library or a gymnastics lesson for pensioners are seen as products to be consumed and where everything is measured in terms of money [...] We do not want to leave them a country in which there are multi-billion Euro rescue packages for ailing banks [...] but nothing for the people with their various social and cultural desires and needs."[44] The squatters occupied the centre for over 111 days and, in the process, they were able to assemble a wide network of support across Berlin and elsewhere in Germany. They even adopted aspects of the action repertoire used by Berlin's longstanding squatting scene. Banners were unfurled and draped from the centre with well-recognised slogans such as "this house is occupied" (*"Dieses Haus ist besetzt"*) and "we will stay" (*"Wir bleiben alle"*) and the group soon became part of a wider network of protests against gentrification and housing scarcity across Berlin. In the face of growing popular support, the local district council agreed, on 18 October 2012, to begin negotiations over the centre's leasehold contract. A one-year temporary use contract was issued though the long-term status of the house remains unresolved (Azozomox, 2014a: 282).

It was the articulation, however, of a right to decent affordable housing in the face of intensifying displacement and dispossession that ultimately played a defining role in the revival of occupation-based practices and the formation of new geographies of protest in Berlin. The return, so to speak, of the 'housing question' was itself hardly surprising. The history of housing inequality in Berlin, after all, has not only depended on recurring cycles of creative destruction, but has also prompted many Berliners to seek alternative forms of housing and shelter. As Friedrich Engels noted in the bloody aftermath of the Paris commune in 1872, the only way capital was able to solve the 'housing question' was to "continually reproduce the question anew" (Engels, 1995 [1872]; see Vasudevan, 2014b). For Berlin, the recrudescence of a *predatory urbanism* has been marked by high rents and housing costs as well as the actual or attempted expulsion of low-income and vulnerable populations from high-value land and locations through gentrification, displacement and forced evictions (Harvey, 2014: 84; see Novak, 2014; Sassen, 2014). As these dynamics have intensified, Berliners have responded through the formation of new tenant organisations and local citizen initiatives which have, in turn, connected with existing activist groups and movements including former squatters, as well as members of the city's well-established autonomist scene and other groups dedicated to anti-racist, anti-imperialist struggles (see Azozomox, 2014a; Holm, 2014a, 2014b).

While the recent emergence of a broad-based movement against housing precarity in Berlin depended, in no small part, on an action repertoire firmly embedded with the city's anti-authoritarian past, it also drew on more recent strategies – creative, experimental, and traversal – that pointed to new forms of "collective, self-articulated protest" beyond the borders of recognised activist spheres (Kanngieser, 2012: 264). Two sets of initiatives, in particular, stand out here and speak to the *making* of new political spaces and horizons in Berlin. The first, the Kotti & Co initiative in the district of Kreuzberg, is a tenants' initiative formed by a group of social housing residents who come from a range of different backgrounds including many from the local German-Turkish community.[45] The second, the 'Coalition Against Forced Evictions' ('*Bündnis gegen Zwangsräumungen*'), is a resistance eviction network that was set up in response to a rapid rise in the number of evictions targeting predominantly low-income tenants across the city (Novak, 2014). Taken together, both cases testify to the persistence of tactics and strategies that seek to challenge and overcome the dominant cartography of the neo-liberal city. As Ceren Türkmen has recently argued, these are developments that point to the articulation of an alternative right to the city that is *actively assembled and shared* rather than simply asserted and claimed (in Novak, 2014: 71).

The origins of the Kotti & Co initiative can be traced back to May 2012 and a street fair on the square in front of Kottbusser Tor which was transformed into a permanent protest camp against rising rents and general housing inequality. The protesters quickly erected a small wooden shack which they described as a '*gecekondu*' in reference to the squatter dwelling that remains a prominent feature of large cities in Turkey (Figure 6.5). In Turkish, *gece* means 'at night' and *kondu* means 'placed' and the term *gecekondu* has comes to describe a house or dwelling 'placed (built) overnight'. According to the social historian and anarchist Colin Ward, the history of the *gecekondu* is, in fact, part of a global history of squatting. This is a history, as I have argued elsewhere, of makeshift rural cottages, precarious and informal urban settlements, experimental housing initiatives and radical autonomous communities (Vasudevan, 2014b). It is a history shaped by a complex patchwork of customary beliefs and rights and epitomised in the widespread view that "if you can build a house between sunset and sunrise, then the owner of the land cannot expel you" (Ward, 2002: 5). If the concept of the 'one-night house' has, according to Ward, "an astonishing global distribution", the illegal erection of a *gecekondu* in a square in Berlin no doubt drew attention to the precarious and often informal practices of dwelling required to 'make do' in urban settings dominated by the logics of capitalist accumulation, most notably, of course, in the cities of the global South (Simone, 2008: 13; see Pieterse, 2008; McFarlane, 2011b; McFarlane and Vasudevan, 2013; Roy, 2011; Vasudevan, 2014b). It also foregrounded the active role that migrant communities have played in the recent history of housing in Berlin and Kreuzberg, in particular, despite

Figure 6.5   The Kotti & Co '*gecekondu*' in Kreuzberg (photo by author).

state-based attempts at "urban governance and control" (Bojadžijev, 2015: 32).[46] These communities were never the passive recipient of official policies and practice. They actively shaped them on their own terms, recalling the dictum famously coined by the architect John Turner who described housing as a process, "a verb rather than a noun" (1976: 62; see Kotti & Co, 2014).

The protest camp set up by the Kotti & Co initiative was not strictly prefigurative. Whilst it became a neighbourhood social centre, it also served as an organising base for a range of different tactics including 'noise demos' (*Lärmdemonstrationen*), community festivals, workshops and conferences. The initiative also set up a Kotti & Co youth organisation. The main aim of the initiative was a repeal of rent hikes imposed by housing associations, including GSW and Hermes, on local tenants living in the large estates that were constructed around Kottbusser Tor as part of the redevelopment of Kreuzberg in the late 1960s and early 1970s described in Chapters 2 and 3. More specifically, the tenants demanded a fixed rate of 4 Euros per square metre and a return of social housing to full public ownership (Holm, 2014a: 13; Kotti & Co, 2014). The Berlin Senate dug in its heels and refused to release any additional funds in order to cap rents. After months of public protests, however, the Senate backpedalled offering a cap of 5.5 Euros per square metre in over 35,000 social housing units across the city. The initiative understandably criticised the Senate's decision as falling far short of their demands and offering little security of tenure to local residents. At the same time, they also recognised their achievement in prompting the Senate to commit new funds to affordable housing in the city. The occupiers have therefore chosen to stay and continue their protest which represented, in their own words, a "struggle for rights (to stay in the neighbourhood, to the city more generally, to be heard and recognised and against structural racism)." "This is not a question," they insisted, "of where and how to perform resistance in opposition to state power at its purest. The people at the Kotti are moved by the question of whether it is possible for those who have no real political voice to affirm a right to protest." As they concluded, "the articulation of a right to the city could never be expressed in parliaments, city streets or courts alone. Rather it is complex self-organized experiences such as the Kotti through which such a right is learned, lived and fought over" (Kotti & Co, 2014: 348, 354).

The re-emergence of a radical housing politics in Berlin has therefore placed particular emphasis on the articulation of a right to the city by, with and on behalf of its precarious, vulnerable and often voiceless inhabitants. In the same year that the Kotti & Co initiative was founded, a new eviction resistance network was established across the city by a coalition of activists involved in housing and other anti-capitalist struggles (Novak, 2014). The 'Coalition Against Forced Evictions' ('*Bündnis gegen Zwangsräumungen*') was formed in the late summer of 2012 and drew immediate inspiration from the PAH anti-eviction movement ('*Plataforma de Afectados por la Hipoteca*' or 'Mortgage Victims' Platform') that was set up in Spain in the wake of the global financial crisis (see Delclós, 2013). In practical terms, the new anti-eviction network has focused its energies on preventing planned evictions through blockades, temporary occupations, demonstrations and other actions that, taken together, constitute a now familiar protest repertoire. It is perhaps no surprise that former squatters and members of various house projects across Berlin have played an active role in the new network. Whilst the main aim of the network is to contest and resist the elemental brutalities of forced eviction, it also works to challenge the sanitising and domesticating discourses normally surrounding 'discardable' people and places and, in doing so, re-centre threatened urban homes as critical terrains of lived and felt habitation, self-organisation and politics.

As a number of scholars have recently argued, forced evictions are destructive processes that hurt, haunt and linger before, during and after their eventuality (see Brickell, forthcoming; Fernández Arrigoitia, 2014). In Berlin, they have resulted in at least one death and the displacement of countless thousands of tenants (Novak, 2014: 43). In response, activists have attempted to generate new forms of solidarity and relation whilst producing oppositional practices that link with and build on similar struggles and movements in other European cities. The past couple of years have also borne witness to a new wave of activist filmmaking in the city as a series of documentaries – including Katrin Rothe's *Beton Gold*

(2013), Gertrud Schulte Westenberg and Matthias Coers's *Mietrebellen* (2014) and Hanna Löwe's *Verdrängung hat viele Gesichter* (2014) – have attempted to connect the experience of forced eviction to the various strategies and tactics of resistance mobilised by ordinary Berliners. And yet, despite these efforts and initiatives, the sheer scale of the police crackdown on eviction resistance protests in Berlin – often involving operations with hundreds of deployed officers – has, in the eyes of many of its residents, only reinforced the view that the city has become the *spatial imprimatur* for a capitalist realism that tolerates few alternatives.[47] If the squatting scene as a broad-based movement was largely pacified in the early 1990s, it is becoming clear that its various afterlives face even greater challenges in developing alternative ways of thinking about and re-imaging the city.

The Autonomous City?

As this chapter has attempted to show, the early 1990s bore witness to the dissolution of the squatting movement in Berlin through a process of legalisation, pacification and co-optation. At the same time, the chapter documents the *re-functioning* of urban squatting and other occupation-based practices by groups of activists and artists as well as a host of other organisations in Berlin from the early 1990s to the present. It is possible, in this context, to discern at least three interrelated developments.

First, that the legalisation of formerly squatted houses provided, on the one hand, a context for new forms of cultural experimentation which mobilised the built form as a site of architectural activism and participatory planning. On the other hand, institutionalisation also pointed to the emergence of new urban policy regimes which attempted to hijack "movement practices for purposes of urban restructuring and enclosure"(Mayer, 2013a: 12; see Colomb, 2012; Novy and Colomb, 2013; Owens, 2008). The enduring legacy of squats as alternative housing projects thus reminded squatters that their survival paradoxically stemmed, in no small part, from the cultural capital they conferred on an increasingly neo-liberal city.

Second, that legalisation also offered many house projects an opportunity – however fragile – to cultivate new forms of self-management and develop collective spaces that explored other identities, intimacies and relations. As the preoccupations of Berlin's activist scene evolved, new political formations also emerged which connected squatting to other networks of protest including anti-globalisation initiatives and increasingly precarious labour regimes as well as anti-racist projects in Berlin and elsewhere. While this meant that squatted spaces were often detached from local geographies of solidarity and engagement, they also established new *transversal* links with other actors at a variety of different scales (see Kanngieser, 2013; Lopez, 2013).

Third, that the intense neo-liberal restructuring of Berlin has precipitated the recent re-emergence of an expanded field of housing activism across the city and the reclamation of a repertoire of contention that depended on the assertion of a right to a very different city. At stake here was the revival of occupation-based practices and other tactics that promoted the assembling of radical urban infrastructures and the development of new practices of shared living that offered "not only inventive ways of perceiving and acting in urban space, but new forms of urban learning and possibility" (McFarlane, 2011b: 182). This was a process that, according to Henri Lefebvre, was contingent on the production of a common field that offered an alternative to the kind of "temporal and spatial shell" solicited by capitalist urbanisation (1991: 384). More than anything else, these practices recognised that the sharp end of neo-liberalism's ongoing process of creative destruction was making life untenable for many of Berlin's more vulnerable inhabitants. New oppositional geographies were therefore produced that attempted to translate the experience of precarity into a politics of radical social change (Vasudevan, 2014b).

It should be clear, therefore, that a strict narrative of movement decline occludes the complex ways in which urban squatting continues to shape the politics of housing in Berlin. It would be equally wrong, however, to romanticise the act of 'occupation'. After all, many political occupations have come and gone. Squats have been repeatedly evicted and other forms of urban protest have been vigorously proscribed while most forced evictions have, in the end, been successful. And yet, while individual examples point to an intensifying urban revanchism, the logic of occupation remains a point of departure for the development of new alternative forms of shared city life. In the end, occupation-based practices may perhaps be best understood as important "laboratories of the politics of the commons" (Feigenbaum et al., 2013: 233). These are laboratories where people have come together to assemble alternative lifeworlds and articulate new forms of contentious politics. These are, moreover, laboratories that depend on an understanding of the city as far more than a mere context for various struggles. Rather, it provides the *substance* – through particular configurations of occupation, self-determination and infrastructure – for the articulation of more just and equal urbanisms (Amin, 2014; McFarlane, 2011b; Vasudevan, 2014b). Whether these initiatives assume a more secure and stable form is the main challenge facing activists and other Berliners for whom the city remains a site of autonomy, emancipation and solidarity.

Notes

1   *taz*, 2.2.2011; *The Guardian*, 2.2.2011.
2   At the time of the eviction, 25 people were living in the house. Only nine were present on the morning of 2 February 2011.
3   *taz*, 3.2.2011; *The Guardian*, 3.2.2011.
4   See http://liebig14.blogsport.de/das-haus/chronik/
5   Hamburger Institut für Sozialforschung (hereafter HIS), "Über die Hausbesetzerbewegung in Ost-Berlin (hereafter HBB), Teil 1", *Telegraph* 9 (1995), pp. 44–45.

6   HIS, "HBB, Teil 1", *Telegraph* 9 (1995), p. 44; see also Robert Havemann Gesellschaft (hereafter RHG), Runde Tische box RTc-01.

7   HIS, "HBB, Teil 1", *Telegraph* 9 (1995). It is worth noting that only 12 of 38 squatted houses in Prenzlauer Berg were represented at the round table meetings. Of the remaining houses, some had already secured individual lease agreements whilst others had expressed a willingness to negotiate. There was also a group of houses that had not confirmed their participation in the negotiation process (letter dated 16.11.1990 from Wohnungsbaugesellschaft to Besetzerrat Prenzlauer Berg, RHG, Runde Tisch file, RTc-01).

8   In January, the Berlin Senate itself collapsed in the wake of the Mainzer Straße evictions.

9   See "Gesprächsrunde über die Ostberliner HausbesetzerInnenbewegung in den 1990er Jahren" (http://www.squatter.w3brigade.de/content/gespraeche/die-strukturen-den-h%C3%A4usern-waren-viel-wichtiger-als-das-haus-sich) (hereafter SQ1). During the negotiations, the squatters were forced to seek legal council. It is in this context that the complex model for securing individual tenancies was first suggested. The former Kreuzberg district councillor Werner Orlowski was also consulted during the process. Orlowski had played an important role in earlier negotiations with squatters in West Berlin in the early 1980s.

10 Papiertiger Archiv (hereafter PTA), typed manuscript, n.d., boxfile, Häuserkampf O-Berlin Mitte.

11 HIS, "HBB, Teil 1", *Telegraph* 9 (1995), pp. 44–45.

12 Houses cleared during this period included Palisadenstraße 49 (26.3.1996); Kleine Hamburger Straße 5 (27.3.96); one wing of Rigaerstraße 80 (11.4.1996); Alt-Sralau 46 (17.4.96); one wing of Kreutziger Straße 11 and Samariter Straße 33 (9.7.1996); Marchstraße/Einstenufer (9.8.96); Linienstasse 158-159 (11.9.96); further eviction of Kreutziger Straße 11,12,13, 21 and one wing of Kinzigstraße 9 (29.10.1996); Niederbarnimstraße 23 (21.5.1997); Scharweberstraße 28, Pfarrstraße 88, part of Schreinerstraße 14 and Rigaer Straße 80 (29.7.1997).

13 *Berliner Morgenpost*, 22.6.92.

14 Stilkam 5 ½ e.V., unpublished MS. Interview with B.H. (March 2009).

15 Stilkam 5 ½ e.V, unpublished MS.

16 Interviews with B.H. and H.G. (February 2009, March 2009, August 2009).

17 *taz*, 21.6.2002; Interviews with B.H. (February, 2009); H.G. (August, 2009).

18 *taz*, 21.06.2002.

19 See http://www.k77.org/

20 SQ1, "Gesprächsrunde über die Ostberliner HausbesetzerInnenbewegung in den 1990er Jahren."

21 In this context, I strongly disagree with a recent commentary by the German autonomist activist Geronimo on the negative effect that feminist politics has played within the squatting scene in Germany (see Geronimo, 2014). Attempts to cultivate safe spaces within squats were a response to serious acts of violence against women and were hardly responsible for policing and undermining the wider 'movement'.

22 The violent sexual assault of a woman who was humiliated, tortured and raped over the course of 12 hours by three other squatters (one man and two women) in a house on Hafenstraße in Hamburg in June 1984 created a storm of controversy that crystallised the debate about sexism and violence within the wider autonomist scene across

West Germany. The fallout prompted many women to seek out 'separatist' alternatives that provided a suitable base for the articulation of autonomist spaces that were ostensibly free from the spectre of patriarchal violence (for a lengthy discussion of the Hafenstraße scandal, see Amantine, 2011; also see Azozomox, 2014b).

23   See http://liebig34.blogsport.de/the-house/house-concept/

24   http://liebig34.blogsport.de/

25   PTA, *BZ* Nr. 15, 5.12.1990, p. 11–12.

26   See PTA, flyer, "Kinzig 9: zu den Ereignisse der letzten Tagen," "The history of Kinzig 9," n.d., Ordner Häuserkämpfe O-Berlin, Friedrichshain; *Kinzig-Bote*, Nr. 0, Oktober, 1998. The history of Kinzig 9 represents something of a microcosm of the wider squatting scene in the post-*Wende* era. The front wing of the house was first occupied in the summer of 1990 and, within a year, it had become well-known with the local drug scene resulting in two deaths in 1991. Over the course of 1991, the dealers living in the house were thrown out by a group of punks and the house quickly became one of the most important meeting-places for the wider punk scene in Germany. At the same time, another group of squatters moved into another wing of the house with a view to establishing a self-organised housing collective. The tensions between the different groups over the future of the house erupted into street fights in August 1992. In the end, the group living in the front wing of the house was evicted in 1996 by the police. Over the course of the next two years, the remaining occupants of the house continued to negotiate with local authorities and were able to secure the house's status as a legal self-organised project in 1998.

27   PTA, *Interim*, 1.10.1998.

28   PTA, *Interim*, 7.9.2000.

29   In 2009, a report by the City of Berlin counted 12 caravan encampments housing over 300 people.

30   http://schwarzerkanal.squat.net/

31   http://de.indymedia.org/2010/03/275163.shtml

32   The title *Umsonst* is derived from the slogan "*Alles für alle, und zwar umsonst* (everything for everyone, and for free)". There were *Umsonst*-related campaigns in a number of German cities including Berlin, Dresden, Freiburg, Hamburg, Köln, Mannheim, München and others. There is a long history of collective appropriation as a radical practice, most notably in the context of Italian *Autonomia*.

33   Renters form the majority of households in major German cities (75 percent in Munich and Cologne, 78 percent in Hamburg and 85 percent in Berlin). See Holm, 2014a: 11.

34   The Hartz reforms are part of a series of reforms to the German labour market undertaken by the SPD government led by Chancellor Gerhard Schröder. The Hartz IV reforms represented the fourth stage of the process and took effect on 1 January 2005. Hartz IV reforms focused on the amalgamation of unemployment benefits with welfare benefits which were recalibrated to the lower level of former social assistance costs (roughly 374 Euros per month for a single person). Housing and health benefits were added on top of this. Recipients were obliged to improve their job prospects or face sanctions levied by the state.

35   The publication of a detailed study on the impact of forced evictions in Berlin appeared as this book was going to press. A comprehensive engagement with the findings of that book was not therefore possible (see Berner, Holm and Jensen, 2015).

36 Berlin is constitutionally obliged to hold a popular referendum on municipal issues through a two-stage process. First, 20,000 signatures are required within a period of six months to request a referendum. Second, a number of signatures equal to or above 7 percent of Berlin's voting population must be secured within a period of four months in order to ensure that the referendum goes ahead. Questions posed in the referendum must correspond to issues which are within the remit and competence of local municipal districts (*Bezirke*).

37 *Der Tagesspiegel*, 14.7.2008.

38 *Berliner Zeitung*, 4.9.2012; *taz*, 2.2.2011; *taz*, 24.11.2009.

39 See http://www.brunnen7.org/; http://linie206verteidigen.blogsport.de/; http://reiche63a.blogsport.de/

40 See http://newyorck.net/

41 *Berliner Zeitung*, 7.6.2005; *taz*, 7.6.2005.

42 *Berliner Zeitung*, 30.6.2012; *The Guardian*, 20.7.2012; *taz*, 29.6.2012.

43 Quoted in *taz*, 9.7.2012.

44 See http://stillestraße10bleibt.blogsport.eu/2012/08/01/aufruf-zum-miteinander/

45 See http://kottiundco.net/english/

46 From 1975 onwards, cities and administrative districts in West Germany were able to ban foreigners from registering in neighbourhoods where they already made up more than 12 percent of the resident population (see Bojadžijev, 2015: 32).

47 The police operation to evict a family from a flat on Lausitzer Straße in Kreuzberg on 14 February 2013 involved over 800 officers (see especially Novak, 2014: 33–42; *Berliner Zeitung*, 14.2.2013; *taz*, 14.2.2013).

Chapter Seven
Conclusion: "Der Kampf geht weiter"

The existential core of urbanism is the desire for radical change.

Edgar Pieterse (2008: 6)

Every image of the past that is not recognized by the present as one of its own concerns threatens to disappear irretrievably.

Walter Benjamin (1969: 255)

On the morning of 26 August 2014, a large contingent of Berlin police officers under the orders of the city's Interior Minister, Frank Henkel, entered into four residential properties that housed over 100 refugees who had been granted temporary asylum in Berlin. The residents were forcibly evacuated though a small group at a hostel on Gürtelstraße were able to evade the police and occupy the roof of the building, threatening to jump in the event of any operation to remove them. The police responded by cordoning off the street and preventing food and other supplies from reaching the rooftop occupiers. At the same time, water and electricity were cut off while supporters began a vigil in front of the police cordon.[1] By early September, little had changed and the situation was becoming critical. As the police tightened their cordon around the occupation, a number of local medical charities demanded that the authorities offer food and medical assistance to the small group of refugees.[2] After 10 days, the occupation finally came to an end as the refugees descended from the roof though they vowed to continue their

Metropolitan Preoccupations: The Spatial Politics of Squatting in Berlin, First Edition.
Alexander Vasudevan.
© 2015 John Wiley & Sons, Ltd. Published 2015 by John Wiley & Sons, Ltd.

protest against a system which, in their view, condemned them to a state of permanent insecurity.

The refugees were part of a larger group of over 550 asylum seekers who had only recently come to an agreement with the Berlin Senate to end a series of protest occupations which had begun in 2012 in response to the 'mandatory residence' (*Residenzpflicht*) requirements affecting foreigners living in Germany, especially applicants for refugee status. According to articles §56 and §85 of the 'German Refugee Processing Law' (*Asylbewerberleistungsgesetz*), applicants are required at all times to reside in the district or *Landkreis* in which they were originally registered. In many cases, refugees were forced to live in camps that were isolated and disconnected from local health and social infrastructures and which were, in many cases, subject to violent attacks by neo-Nazi groups. As they awaited a decision on their status, often for months or even years, applicants received a small stipend from the state, vouchers for food and could work for the meagre sum of 1.05 Euros per hour.[3]

In March 2012, the suicide of an asylum seeker in a camp in the Bavarian city of Würzburg was the catalyst for a wave of nationwide protests that began with a hunger strike and culminated in a 600km march from Würzburg to Berlin in September 2012 by a group of refugees and their supporters. The refugees framed their actions as a 'strike' against the various legal proscriptions that shaped and structured their everyday lives. We are protesting, they proclaimed, "to gain our rights and to defend our right to an honourable life." The 'strikers' demanded a swift end to the current process of adjudication that left many of them in a state of legal limbo, the closure of the system of camps in which they were housed and the right to freedom of movement within Germany.[4] At the same time, a series of improvised protest camps were set up in cities across Germany in solidarity with the marchers, the most important of which was located in Berlin on Heinrichplatz in the district of Kreuzberg.

After 28 days, the marchers reached Berlin and soon joined with the occupiers on Heinrichplatz to form a new protest camp at nearby Oranienplatz. The occupation was followed, a week later, by the largest-ever demonstration in support of refugees and asylum seekers in Germany.[5] As the camp at Oranienplatz grew in size, the protesters and their supporters adopted a form of protest that drew on a familiar range of strategies which owed their provenance, in no small part, to the long history of anti-authoritarian revolt in Germany (Brown, 2013; Klimke, 2010; Slobodian, 2012). This was a history, as this book has argued, that depended on the built environment and, more generally, geography for the articulation of a repertoire of contention that depended, in turn, on the production of new practices and solidarities that were firmly rooted in an understanding of urban space as a key site of action and revolt. At stake here, in other words, was a *spatial politics* that transformed German cities into veritable theatres of protest (see Davis, 2008; Vasudevan, 2011a).

For the 'striking' refugees, it was the impulse to occupy and reclaim space as a tool of protest and resistance that became a key feature of their actions (see Vasudevan, 2014a). Alongside the protest camp at Oranienplatz, a host of other occupation-based tactics were mobilised in conjunction with a number of activist groups in Berlin. In December 2012, the former Gerhard Hauptmann School on Ohlauer Straße in Berlin-Kreuzberg was squatted by refugees in conjunction with a group of housing activists and anti-gentrification protesters and with a view to forming a 'project house' and a community social centre. The Refugee Strike House and the Irving Zola House were, in this way, established as a base for a number of initiatives within the local neighbourhood (Azozomox, 2014a).[6] The occupants of the former school called for the formation of a centre that offered full and free accessibility to all Berliners. In an open letter to local authorities, the occupiers argued that there was a serious shortage of free self-managed spaces across the city. "In a neighbourhood," they added, "where rents are not only rising but tenants are threatened with forced eviction [...], a barrier-free and accessible social centre is long overdue."[7]

The protest camp at Oranienplatz was cleared in April 2014 by the police, while the occupants of the squatted school agreed in July 2014 to evacuate the centre in exchange for basic accommodation and financial support. At the same time, an agreement was reached to transfer to Berlin the cases of over 550 asylum seekers, many of whom were directly involved in the two occupations. Whilst it was promised that the cases would be reviewed impartially, local authorities later reneged on their original commitments and, in late August 2014, 108 of the original 550 refugees received orders to either leave Berlin for the districts in which they were registered or return to their original EU country of entry.[8] Housing and financial support was also withdrawn which ultimately led to the police operation on 26 August 2014 and the rooftop occupation of the hostel on Gürtelstraße.

For the refugees, their putative *illegalisation* was part of a wider constellation of policies that served as a refusal in the case of the German state to grant migrants and refugees the right to reside and work legally in the country. It would be tempting in this respect to detect in their experience an instantiation of the 'naked' or 'bare' life described by the philosopher Giorgio Agamben as a life stripped of political inscription insofar as the asylum seeker or refugee is excluded from the protective order of the state (1993; see also Agamben, 1998; Slobodian, 2013a).[9] And yet, far from a return to a prepolitical biological order, these are, infact, lives produced and socially saturated by power (Chamayou, 2012: 140). As Judith Butler has argued in a related context, modern 'illegals' are not "undifferentiated instances of 'bare life' but highly juridified states of dispossession" (in Butler and Spivak, 2007: 42). These are states of exclusion that are, in turn, *spatially generative* insofar as they produce the kind of grey spaces (protest camps, squats, etc.) recently described by Ananya Roy and Oren Yiftachel that hover – precariously – on the border between legality and approval and eviction and destruction (Roy, 2011: 235; Yiftachel, 2009: 89).

The various occupations undertaken by the striking refugees in Berlin may plausibly be seen, therefore, as the articulation of an alternative right to the city that is, at the same time, an expression of a basic fundamental right to *be in the city*. These are actions that, on the one hand, imagine a right to participate in the production of urban space. On the other hand, they also anticipate and prefigure other forms of care, generosity and dwelling whose history is unthinkable without envisioning and understanding Berlin as a city of migration (Kotti & Co, 2014; see Bojadžijev, 2008; Borneman, 1992; Hinze, 2013; Karakayali, 2009; Mandel, 2008). It is in this very context that the recent wave of refugee protests have come to represent just the latest episode in an expansive geography of occupation that is the main subject of this book. While these are protests that seem far removed from the squatting of dilapidated tenement blocks in the late 1970s and early 1980s, the solidarities and connections recently developed between refugees, former squatters and other housing activists point to the enduring significance of occupation as a set of tactics for how we might still come to know and live the city differently.

It is this very basic question about the relationship between the history of squatting in Berlin and the *making* of alternative urbanisms that is at the centre of this book. As I have argued, this is a book about the historical development of the squatter movement in Berlin. It is also a book about the challenges that we face in writing histories of radical urban politics and their relationship to the development of more just and equal spaces in our cities. As an exercise in spatial history, the book sets out to do two things. First, it offers a detailed historical reconstruction of Berlin's squatting milieu from the late 1960s to the present, focusing on what squatters actually *did*, the terms and tactics they deployed, the ideas and spaces they created. Second, it makes a broader case for the importance of squatting to recent scholarship on the long and complex history of the New Left in Germany and its various afterlives (Brown, 2013; Davis, 2008; Davis et al., 2010; Hannah, 2010; Klimke, 2010; Reichardt, 2014; Slobodian, 2012; Thomas, 2003; Varon, 2004). Taken together, the book challenges accounts that relegate the history of squatting in Berlin and West Germany to a strict supporting role within an anti-authoritarian revolt shaped by a "few trademarked representatives or iconic leaders" (Rethmann, 2011: 46; see also Ross, 2002). It also resists attempts to 'capture' and re-cast the historical legacy of the New Left as a definitive moment in the 'fundamental liberalisation' of West German society (Habermas, 1988; see Brand et al., 1986).

To do so, the book locates the genesis of the squatter movement in Berlin within a broader history of capitalist accumulation and uneven urban development. While it would be wrong to conflate the practices of squatters in Berlin with a much earlier repertoire of contention, they do nevertheless form part of an expansive field of dissent and resistance. Squatting represented, according to this view, the political *other* to 'creative destruction' and we find in the various lives, spaces and practices of squatters an alternative urban imagination that transformed Berlin into a makeshift

Figure 7.1   Montage of the West Berlin Squatter Scene in the *Instand-Besetzer-Post*, 30.4.1981, n.p. (Papiertiger Archiv Berlin).

laboratory for developing new forms of collective living (Figure 7.1). At the same time, the book also adopts an approach that seeks to remain alert to the often neglected heterogeneity of the movement – its various approaches, challenges and goals – by taking in developments before and after the fall of the Berlin Wall and in both the former West and East of Berlin. There are, of course, risks in eschewing periodisations that have yielded important insights into the history of the New Left in Germany just as the fulsome emergence of new social movements in the 1960s and 1970s in West Germany should not be seen as somehow homologous to the tentative development of an alternative public sphere in the final years of the German Democratic Republic. And yet, at stake here, if anything, is a rather different concern for marking the fractured and uneven emergence of squatting in Berlin and elsewhere in Germany as a set of practices that were closely connected to the anti-authoritarian revolt of the late 1960s but that also exceeded and extended the contours of that history, producing new radical lines of flight for which the city remains a key site of action and dissensus.

This book began, in this way, with a series of basic questions. Why did thousands of citizens choose to break the law and occupy empty flats and other buildings in numerous cities across Germany and in Berlin in particular? Were these actions dictated by pure necessity or did they represent a newfound desire to imagine other ways of living together? Who, it asks, were these squatters? And what were the central characteristics of urban squatting (goals, action repertoires, political influences)? In order to answer these framing questions, I have argued that the history of squatting in Berlin can only adequately be understood *geographically*; that is, as a series of spatial practices aimed at creating alternative livable places that are themselves tied to particular forms of activism and empowerment. As I have shown, the widespread suppression of the extra-parliamentary opposition in West Germany during the 1960s and 1970s prompted activists to seek out, in the words of the German filmmaker Alexander Kluge, "new programs of existence" and experiment with practices and tactics including squatting and site occupation (1973: 5).[10] But more than this, the composition of such oppositional geographies also promoted a radical understanding of the urban itself through the development and circulation of new activist imaginations and the production and performance of collective modes of living that attempted – sometimes successfully, sometimes less so – to exist beyond dominant power relations and redefine what it means to live in a city.

As a *thick description* of the history of the squatting scene in Berlin, the preceding chapters have placed a particular emphasis on documenting the everyday experiences, practices and sentiments of squatters and the role that they have come to play in the making of radical political horizons. The book's theoretical footprint is, in this respect, conspicuously modest and keyed to exploring the relationship between squatting as an urban social movement and the possibilities for building and sustaining alternative political imaginaries. It is not my intention, in other words, to develop a theory of urban squatting that is all-encompassing. Rather, my own aim has been to hold onto an optic that zooms in on the spatial politics of squatting while attending to Colin McFarlane's recent appeal for a "different theorization and lexicon of urbanism that seeks not to displace existing urban theory, but to add to it" (2011b: 184; emphasis added). It is with this squarely in mind that I offer this book as a response to the question recently posed by David Harvey (2012: 120) as to whether struggles within and over the city, and over the qualities and prospects of urban living, should be seen as fundamental to anti-capitalist politics. While there is no easy answer to this question, the *normative* demand for an 'alternative city' is not confined to Berlin and has, if anything, become an increasingly pressing issue in an age of austerity dominated by new forms of displacement, marginalisation and privatisation (see Peck, 2013a).

In the preceding pages, I have therefore developed an empirically grounded account of the history of the squatter movement in Berlin that also sets out to generate an original geographical framework for investigating and rethinking the possibilities of radical urban change. To do so, I have drawn on existing historical scholarship on the development and constitution of urban social movements in Germany. I have also turned to wider geographical debates about the production of alternative urbanisms whilst engaging with a range of cultural theory and radical political philosophy that includes *inter alia* the work of Michel Foucault, Félix Guattari, Hans-Jürgen Krahl, Herbert Marcuse and Antonio Negri. The modest theoretical focus at the heart of this book is, in this way, contingent on concepts and debates that were *immanent* to the practices of the countercultural Left in Germany and elsewhere in Europe as they emerged in the 1960s and 1970s. If concepts, as Gilles Deleuze and Félix Guattari once argued, must be "invented" or "created", it is entirely plausible – even desirable, in my view – to see in the makeshift practices of squatters ways of thinking about and living urban life that have the potential to reanimate the city as a key site of geographical inquiry (Deleuze and Guattari, 1994: 5; see Vasudevan, 2014b).

As a number of scholars have recently argued, any adequate theory of squatting and its various geographies needs to "get to grips with the fundamental ambiguity that often characterises life in these places" (see Pithouse, 2006). The conceptual armature required must, to a large extent, accommodate and inhabit the very provisionality of squatting itself. This is, moreover, a form of conceptualisation that demands a heightened sensitivity to those aspects of city life that cannot be simply captured or be fully expressed by the traditional categories of modern urban geography. In the case of Berlin, a close engagement with the ambiguity and inventiveness of squatting – from the actions of citizens' initiatives in Kreuzberg to the practices of *Schwarzwohner* in Prenzlauer Berg, from the experimental geographies of artists to the protest camps of anti-gentrification activists – has at least three broader implications for how we theorise the urban as a staging ground for new capacities and potentials. First, it highlights the need for a committed empirical approach that places particular emphasis on the *making* of alternative urban spaces and infrastructures. Second, it identifies the key role that *emotional labour* plays in transforming radical political goals into shared spaces of action and solidarity, care and generosity. Third, it calls for an *historical perspective* that re-imagines the city as a living archive of alternative knowledges, materials and resources.

Throughout this book, I have focused on the everyday practices and lived materialities of squatting as a *makeshift urbanism* through which precarious lifeworlds were pieced together and sustained (see also McFarlane and Vasudevan, 2013; Vasudevan, 2014b). After all, squatters in Berlin often confronted abandoned or dilapidated spaces that required significant renovation and, for many, the very act of occupation was understood as a form of 'resuscitation'. Squatters thus cultivated an ethos of self-determination and autonomy – a radical DIY empiricism – that focused on the rehabilitation of buildings and the active assembling of new forms of dwelling. In practical terms, this depended on a modest ontology of mending and repair that brought people together with ideas, practices, resources and things. A preoccupation with the active assembling of radical political spaces prompted squatters, on the one hand, to develop and share new forms of 'urban learning' that were centred on the spaces in which they lived and worked and 'the practices of habitation' they produced (McFarlane, 2011b; Lefebvre, 2014[1976]). On the other hand, the actions of squatters were equally tied to a broader struggle to re-imagine city life as a shared political project that connected the micropolitics of occupation to the cultivation of new geographies of action and solidarity. As a *radical politics of infrastructure*, squatting represented far more than a discrete act of rehabilitation and repair. It also encompassed a wider set of processes that were instrumental in the articulation of an alternative right to the city embedded within local neighbourhoods and connected to wider ecologies of protest and resistance.

Infrastructure, as AbdouMaliq Simone reminds us, is "commonly understood in physical terms, as reticulated systems of highways, pipes, wires, or cables". For Simone, these modes of provisioning and articulation are often viewed as playing a central role in the reproduction of urban life (2004: 407). Simone, whose own work is firmly embedded within the cities of the global South, argues that we need to *extend* our notion of infrastructure to people and their capacity "to generate concrete acts and contexts of social collaboration" (2004: 419). While the activities and ideas of squatters in Berlin should not be equated with the experience of residents in African cities, it is nevertheless possible, as I have argued elsewhere, to spy connectivities across multiple sites that seek to open up a research platform for rethinking how different experiences of urban precariousness might be shared across the North/South divide (Vasudevan, 2014b; Hentschel, forthcoming; Roy, 2011). For members of Berlin's squatter movement, it was the reclamation of the built environment and its various infrastructures that ultimately served as a radical platform for the constitution of other ways of being in the city. It is against this backdrop, furthermore, that German squatters adopted the motto *Instands(be)setzung* as a slogan for the movement – the term itself a combination of the German for maintenance (*Instandsetzung*) and squatting (*Besetzung*). To the extent that squatted spaces and their occupants were therefore able to build a set of infrastructures related to their aims as sites of autonomy and self-determination, repair and inhabitation, these *material geographies* also played, as I have suggested, a central role in the formulation of an alternative urban politics. The problem of infrastructure represented, in this way, an important point of departure for how squatters understood their own actions as a spatial politics and as a form of "infra-commoning" to borrow Ash Amin's felicitous phrasing (2014).

It would, of course, be misleading to advance a dispassionate view of the history of squatting in Berlin untainted by dissent and disagreement. The action repertoire adopted by squatters in the city often varied from house to house and there were, as this book has demonstrated, significant differences and discontinuities across a long and uneven history of protest and resistance. The ability to act as a *movement* was not only dependent on the mobilisation of specific material practices and skills. It also enrolled a range of emotional practices and affective states that played an important role in sustaining new spaces, infrastructures and networks. "Emotion," as Deborah Gould (2009) reminds us, "is fundamental to political life." It can constrict, she adds, "our political imaginaries [...] as well as extend them in new, unexpected directions" (439, 443). Berlin's squats were, in this context, spaces saturated with intense feelings that ranged from joy and hope to anger and despair as squatters were moved by their actions and accomplishments and correspondingly shaken by the losses and failures they experienced. At the same time, these were also affective spaces governed by passions and sensations – visceral, bodily, sensory and nonconscious – that were difficult to describe but were nevertheless able to stir new political formations (see Feigenbaum et al., 2013: 20).

To be sure, the emotional and affective resonances of squatting were never simply an unintended effect or consequence of actions, decisions and interventions undertaken by individuals or groups. As a growing number of scholars have recently argued, the very emergence of the New Left in Germany in the late 1960s was characterised by a critique of capitalism that focused on the 'emotional regime' that it allegedly produced (Reddy, 2001; see especially Boyle, 2012; Häberlen and Smith, 2014; Reichardt, 2014; Slobodian, 2012). As I have argued, this was a critique developed by squatters as well as other radical groupings that connected the production of new emotional practices to oppositional geographies of care, connection and solidarity. For many, what was urgently needed was a new 'warmth' and empathy that actively challenged the alienation and anonymity of an organised, technocratic society (see Sennett, 1970). As two Berlin activists writing in 1981 at the very height of the squatter movement concluded, "based on our subjective experiences and needs and repelled by an all-encompassing coldness, mendacity and aimlessness, we realised that one needed to turn away [...] We knew that one could not longer live as we used to" (Bacia and Scherer, 1981: 20). "We want to build our own blissful island in the middle of Kreuzberg" were the words of another contemporary publication (quoted in Lang, 1998: 145).

If the aspirations, dreams and wishes of many squatters and their supporters were shaped by appeals to a new 'innerness' [*Innerlichkeit*] and a turn to settings that offered the prospect of "solidarity, warmth and tenderness," these were feelings and impulses that were influenced by many factors that both generated and restricted a sense of political possibility (Glaser, 2000: 371; Mettke quoted in Reichardt, 2014: 198; Gould, 2009: 440; see also Plowman, 1998). Large-scale economic shifts, altered geopolitical realities, and various movement defeats all played a significant role in shaping the emotional habitus of squatting from the late 1960s to the present. While this was a history and geography that ultimately failed to meet the full demands and desires of its makers, it was responsible for the emergence of a vast network of radical spaces in Berlin and elsewhere in Germany – cafés, pubs, alternative presses, bookstores, youth centres, and squats – in which friendships were made, traditional identities and intimacies challenged and solidarities secured. Many of these spaces have since disappeared though there remains an awareness and widespread appreciation of the *emotional labour* that went into their making.

A focus on the emotionally saturated nature of contentious politics draws attention to the highly charged ways in which an alternative urbanism and the right to a different city are often imagined and understood. A recognition of the attachments, disagreements and pleasures that underwrite the formation of radical urban lifeworlds cannot, however, be dissociated from the broader political projects they embody. A key argument of this book is that a close reading of the micropolitical tactics adopted by squatters must also be framed as a critical disquisition on the articulation of alternatives to uneven geographical development and the urbanisation of capital (SQEK, 2014; see also Pruijt, 2013; Vasudevan, 2014a). It is also for this reason that this book assumes the form of a spatial history that seeks to remain alert to the rich sedimented history of practices and imaginaries that transformed Berlin into a crucible of protest from the late 1960s onwards. If the book offers, in the end, a new vantage point on the history and geography of the New Left in Germany, it does not speak to a blithe historicism. Rather a scholarly preoccupation with locating and registering the vast archive of practices mobilised by squatters in Berlin is shaped, if anything, by a commitment to marking their enduring significance for the development of a radical urban geography. My aim here is to acknowledge the possibilities and challenges that have accompanied the recent revival of squatting and other occupation-based forms of resistance across a new transnational landscape of protest and dissent. The occupation and reclaiming of urban public space and the assembling of improvised protest camps became, after all, a defining image of the Arab Spring and the Indignados Movement in southern Europe, while the recrudescence of a global austerity urbanism has also led to new forms of squatting in response to intensifying forms of housing precarity (Mitchell, 2012; Sevilla-Buitrago, 2011; SQEK, 2013, 2014; Vasudevan, 2014a). This has, of course, produced a differential geography of practices and experiences across the Global North and South. At stake here for many, as Judith Butler (2012) has argued, is an incipient right to produce a different world that "questions structural inequality, capitalism, and the specific sites and practices that exemplify the relation between capitalism and structural inequality" (11). For others, the very space of occupation became a key site through which an emancipatory urban politics was articulated and developed, often in opposition to specific forms of development and displacement. And for others still, the occupation and reconfiguration of public space was a simple demonstration of their basic right to be and persist in that very space.

It is with all of this in mind that I have offered, in these pages, an account of the history of squatting in Berlin that should also be read, more generally, as an archaeology of our present and as a powerful critique of the revanchist logics of neo-liberal urbanisation. I do not mean to romanticise or fetishise this history nor is it my intention to develop a critical geography of occupation that serves as a defining model for how a radical urban politics should be conducted. As Margit Mayer and a number of other scholars have pointed out, neo-liberal urban policies have successfully captured and instrumentalised dynamic local subcultures as part of a strategy of cultural revitalisation and creativity-led economic and urban development (2013b: 4; see Balaban, 2011; Colomb, 2012; Peck, 2012; Uitermark, 2004, 2011). In this context, squatted spaces and social centres are often transformed into 'branding assets' that contribute to the refunctioning of cities as creative experimental sites attractive to economic investment and further marketisation. Such processes can be observed in many European and North American cities, including Berlin, and have only accelerated in the wake of the financial crisis which has, according to Mayer, "provided a rationale for dismantling alternative infrastructures and for cutting back on funding for self-organised projects of all kinds – at the same time as neoliberal urban regeneration and hyper-gentrification of central city land accelerates displacement pressures and threatens leftist 'free zones'" (2013b: 5).

In the end, Mayer is right to argue that contemporary struggles over squatting and other forms of radical housing-based activism are taking on a "qualitatively new political significance" as a constellation of practices that have the potential to disrupt and undermine the urbanisation of capital (2013b: 6). This is a struggle that highlights, as I have argued, the central role of geography, more generally, in the struggle for social justice. I want to close the book, however, by suggesting that this is also a struggle over the very meaning and identity of Berlin itself. As much as the recent rebranding of the city has depended on the colonisation and pacification of its various radical histories, the recent emergence of new oppositional movements in the city from struggles over refugee rights to protests over gentrification, rising rents and forced evictions show that these histories can still speak to other ways of thinking about and inhabiting the city that offer a meaningful, wide-ranging alternative to the predations of capital. I write these concluding words, therefore, in the knowledge that a church at Mariannenplatz in Kreuzberg has just been squatted by a group of refugees whose ongoing struggles were the subject of this chapter's opening pages.[11] The St.-Thomas-Kirche lies only 100 metres away from the very first squat in Berlin, the Georg von Rauch-Haus, which was occupied in December 1971 and remains an alternative house project and iconic site within Berlin's radical scene. While the fortunes of the city's squatter movement have since waxed and waned, it is clear that the spaces they created and the practices they deployed still resonate for many activists (and former squatters) for whom a more socially just urbanism persists as an unfinished project.

This is, admittedly, an increasingly fragile project as the few autonomous spaces that have survived in Berlin, either as legalised house projects or community social centres, are now increasingly under threat by the city's baleful neo-liberalisation (Azozomox, 2014a; Holm, 2014a, 2014b). The concomitant ability of squatters and other housing activists to scale up their actions as part of a wider process of commoning linked to new practices of connection and inhabitation remains an open question. These developments only reinforce the importance of retracing the history of squatting in Berlin gathering together the various imaginings, practices, artefacts and words that form its archive. In this respect, I am reminded of a passage in Nanni Balestrini's novel *The Unseen* which explores the emergence of the militant movement known as *Autonomia* in Italy in the 1970s through the story of a single working-class protagonist from high-school rebellion through to his arrest and incarceration. In one arresting passage, the protagonist reflects on the confiscation and destruction of his own personal movement archive by the police:

> I wasn't worried about the search because I knew there was nothing incriminating in the house the only thing that could interest them was in the cellar my records of the movement all the newspapers the magazines the leaflets of these past years I was jealous of my archive I'd spent hours organising it but it was all perfectly legal…and so they started the gradual removal from the cupboard to the boots of cars I was wretched I knew I'd never see it again it would rot in the cellars of some police station or court house it would vanish just as in years to come all the comrades' archives would vanish deliberately destroyed by them all the newspapers all the magazines all the leaflets all the documents all the posters all the publications of the movement destroyed vanished all bundled in cardboard boxes and plastic rubbish bags and burned or thrown on rubbish tips tons of printed matter the written history of the movement its memory dumped among refuse consigned to the flames (Balestrini, 2012: 86–87).

Unlike the case of Balestrini's militant protagonist, the archives assembled by Berlin's squatters do still survive offering, in the words of Michel Foucault, a "precarious domicile of […] words" far more durable perhaps than the spaces they described (Foucault, 2000: 162). If this book has attempted to trace, inhabit and re-animate these spaces in the hope that their stories may point to a radically different understanding of shared city life, be it in Berlin or elsewhere, it ultimately concludes with the realisation that doing justice to the dreams and desires of squatters in the face of their criminalisation and confiscation is itself an increasingly radical act. It with this in mind that I offer this book – the stories it tells, the spaces it revisits, the people it follows, the practices it documents – as a modest reminder that the struggle for an alternative right to the city really matters and that this is a struggle that continues.

Notes

1 *Der Tagesspiegel*, 27.8.14; *taz*, 30.8.14.
2 *taz*, 5.9.14; *Neues Deutschland*, 5.9.14 http://guertelstraße.wordpress.com/2014/09/06/besuch-auf-henkels-sommerfest-und-arztinnen-die-kein-wasser-zum-dach-bringen-durften/
3 See http://asylstrikeberlin.files.wordpress.com/
4 http://asylstrikeberlin.files.wordpress.com/
5 *taz*, 14.10.2012.
6 Irwing Zola (1935–1994) was an American activist in the field of medical sociology and disability rights.

7   See http://irvingzolahaus.blogsport.de/images/offener_brief_soziales_zentrum.pdf

8   *Junge Welt*, 3.9.2014.

9   The historian Quinn Slobodian (2013a) has recently argued that the 1965 'Foreigner Law' (*Ausländergesetz*) provided an extra-legal basis for circumventing the constitutional rights of foreigners living in West Germany. Slobodian also shows how activist groups in the 1970s mobilised left-legal constitutional principles to defend these rights and challenge the exercise of police power.

10 Kluge, it should be said, was sceptical of the New Left's ability to persist in a programme of revolutionary transformation despite his interest (alongside) Oskar Negt in developing a "microphysics of resistance" (see Kluge and Negt, 2014 [1981]).

11 *taz*, 13.9.14.

References

Archives and Archival Collections

Archiv, "APO und soziale Bewegungen", Freie-Universität Berlin (APO-A)
Archiv, Hamburger Institut für Sozialforschung, Hamburg (HIS)
Kreuzberg Museum, Berlin (KM)
Landesarchiv Berlin (LAB)
Papiertiger Archiv, Berlin (PTA)
Robert-Havemann-Gesellschaft, Archiv, Berlin (RHG)

Magazines and Periodicals

Agit 883
BZ
InfoBUG
Instand-Besetzer-Post
Interim
Kursbuch
Linkeck
Radikal
Das Schwarze Kanal
Südost Express
TIP
Wir wollen alles

Metropolitan Preoccupations: The Spatial Politics of Squatting in Berlin, First Edition.
Alexander Vasudevan.
© 2015 John Wiley & Sons, Ltd. Published 2015 by John Wiley & Sons, Ltd.

Newspapers

Berliner Morgenpost
Berliner Zeitung
Der Abend
Deutsche Allgemeine Zeitung
Lokalanzeiger
Neues Deutschland
Die Neue preussische Kreuzzeitung
Norddeutsche Allgemeine Zeitung
Spandauer Volksblatt
Spenersche Zeitung
Die Rote Fahne
Der Spiegel
Der Tagesspiegel
die tageszeitung
The Guardian
Vossische Zeitung
Die Zeit

Secondary Sources

Aalbers, M. and Christophers, B. 2014. "Centering Housing in Political Economy." *Housing, Theory and Society*, 31: 373–394. DOI: 10.1080/14036096.2014.947082.

Adey, P. and Kraftl, P. 2008. "Architecture/affect/inhabitation." *Annals of the Association of American Geographers*, 98: 213–231. DOI: 10.1080/00045600701734687.

Agamben, G. 1993. *The Coming Community*. Minneapolis, MN: University of Minnesota Press.

Agamben, G. 1998. *Homo Sacer: Sovereign Power and Bare Life*. Stanford, CA: Stanford University Press.

A. G. Grauwacke. 2003. *Autonome in Bewegung: Aus den ersten 23. Jahren*. Berlin: Assoziation A.

Agnoli, J. 1979. "Jesuiten, Kommunisten und Indianer." In *Zwei Kulturen: TUNIX, Mescalero und die Folgen*, edited by D. Hoffmann-Axthelm et al. Berlin: Verlag Ästhetik und Kommunikation, pp. 80–93.

Ahmed, S. 2007. *Queer Phenomenology: Orientations, Objects, Others*. Durham, NC: Duke University Press.

Aly, G. 2008. *Unser Kampf: 1968 – ein irritierter Blick zurück*. Frankfurt/M: Fischer Verlag.

Amantine. 2011. *Gender und Häuserkampf*. Münster: Unrast Verlag.

Amantine, ed. 2012. *"Die Häuser denen, die drin wohnen! Kleine Geschichte der Häuserkämpfe in Deutschland."* Münster: Unrast Verlag.

Amin, A. 2013. "Seeing like a city." Talk at the University of Nottingham, UK, 2 May.

Amin, A. 2014. "Lively Infrastructure." *Theory, Culture & Society*, 31: 137–161. DOI: 10.1177/0263276414548490.

Anon. 1981. "No title." *Stadtzeitung für Freiburg*, April 1981: 41–42.

Arendt, H. 1973. *The Origins of Totalitarianism*. New York, NY: Harcourt, Brace, Jovanovich.

Arndt, S. et al., eds. 1992. *Berliner, Mainzer Straße*. Berlin: Basisdruck.

Arps, J.O. 2011. *Frühschicht: Linke Fabrikinterventionen in den 70er Jahren*. Berlin: Assoziation A.

Attoh, K. 2011. "What Kind of Right is the Right to the City?" *Progress in Human Geography*, 35: 669–685. DOI: 10.1177/0309132510394706.

Aust, S. 1985. *Der Baader Meinhof Komplex*. Hamburg: Hoffmann & Campe.

Aust, S. and Winkler, W. 2008. *Die Geschichte der RAF*. Hamburg: Rowohlt.

Azozomox. 2014a. "Besetzen im 21. Jahrhundert 'Die Häuser denen, die drin wohnen.'" In *Reclaim Berlin: Soziale Kämpfe in der neoliberalen Stadt*, edited by A. Holm. Berlin: Assoziation A, pp. 273–304.

Azozomox. 2014b. "Squatting and Diversity: Gender and Patriarchy in Berlin, Madrid and Barcelona." In *The Squatters' Movement in Europe: Commons and Autonomy as Alternative to Capitalism*, edited by SqEK. London: PlutoBooks, pp. 189–210.

Azozomox. n.d. "Squatting in Berlin, 1970–2014." Working paper.

Bacia, J. and Scherer, K-H., eds. 1981. *Paßt bloß auf! – Was will die neue Jugendbewegung*. Berlin: Olle Verlag.

Bailey, R. 1973. *The Squatters*. Harmondsworth: Penguin Books.

Balaban, U. 2011. "The Enclosure of Urban Space and Consolidation of the Capitalist Land Regime in Turkish Cities." *Urban Studies*, 48: 2162–2179. DOI: 10.1177/0042098010380958.

Balestrini, N. 2012. *The Unseen*. London: Verso.

Bashore, J. 1991. *The Battle of Tuntenhaus*. Video, 25 min.

Bashore, J. 1992. *The Battle of Tuntenhaus*, update. Video, 20 min.

Baumann, C. et al., eds. 2011. *Links alternative Milieus und Neue Soziale Bewegungen in den 1970er Jahren*. Heidelberg: Winter Verlag.

Beck, J. et al., eds. 1975. *"Jetzt reden wir." Betroffene des Märkischen Viertels*. Hamburg: Rowohlt.

Becker, J. 1977. *Hitler's Children: The Story of the Baader-Meinhof Terrorist Gang*. Philadelphia: Lippincott.

Beier, R. 1982. "Leben in der Mietskaserne: Zum Alltag Berliner Unterschichtsfamilien in den Jahren 1900 bis 1920." In *Hinterhof, Keller und Mansarde: Einblicke in Berliner Wohnungselend 1901–1920*, edited by G. Asmus. Hamburg: Rowohlt, pp. 244–270.

Benjamin, W. 1969. "Theses on the Philosophy of History." In W. Benjamin, *Illuminations*. New York: Schocken Books, pp. 253–264.

Benjamin, W. 1974–1991. *Gesammelte Schriften*. 7 vol. Frankfurt/M: Suhrkamp Verlag.

Benjamin, W. 1999. *The Arcades Project*. Cambridge, MA: The Belkap Press.

Benjamin, W. 2014 [1930]. "The Rental Barracks." In W. Benjamin, *Radio Benjamin*. London: Verso, pp. 56–62.

Berdahl, D. 1999. *Where the World Ended: Re-unification and Identity in the German Borderland*. Berkeley, CA: University of California Press.

Berlant, L. 2011. *Cruel Optimism*. Durham, NC: Duke University Press.

Berndt, H. 1968. *Das Gesellschaftsbild bei Stadtplanern*. Stuttgart: Karl Krämer Verlag.

Berndt, H. 1969. "Kommune und Familie." *Kursbuch*, 17: 129–146.

Berner, L., Holm, A. and Jensen, I. 2015. *Zwangsräumungen und die Krise: Des Hilfesystems*. Berlin: Institut für Sozialwissenschaften.

Bernet, C. 2004. "The "Hobrecht Plan" (1862) and Berlin's Urban Structure." *Urban History*, 31: 400–419. DOI: 10.1017/S0963926805002622.

Bernstein, E. 1907. *Die Geschichte der Berliner Arbeiter-Bewegung: Ein Kapitel zur Geschichte der deutschen Sozialdemokratie.* Berlin: Buchhandlung Vorwärts.

Bernt, M. and Holm, A. 2009. "Is It or Is It Not? The Conceptualisation of Gentrification and Displacement and its Political Implications in the Case of Prenzlauer Berg." *City*, 13: 312–324. DOI: 10.1080/13604810902982268.

Betts, P. 2008. "Building Socialism at Home: The Case of East German Interiors." In *Socialist Modern: East German Everyday Culture and Politics*, edited by K. Pence and P. Betts. Ann Arbor, MN: The University of Michigan Press, pp. 96–132.

Betts, P. 2010. *Within Walls: Private Life in the German Democratic Republic.* Oxford: Oxford University Press.

von Beyme, K. 1987. *Der Wiederaufbau: Architektur und Städtebaupolitik in beiden deutschen Staaten.* Munich: Piper.

Bezirksamt Kreuzberg. 1956. *Wir bauen die neue Stadt: Die städtebauliche Neugestaltung der Luisenstadt im Bezirk Kreuzberg.* Berlin: Bezirksamt Kreuzberg.

Biehl, J. 2005. *Vita: Life in a Zone of Social Abandonment.* Berkeley, CA: University of California Press.

Bieri, S. 2012. *Vom Häuserkampf zu neuen urbanen Lebensformen: Städtischen Bewegungen der 1980er Jahre aus einer raumtheoretischen Perspektive.* Bielefeld: Transcript Verlag.

Birke, P. and Larsen, C., eds. 2007. *Besetze deine Stadt: Häuserkämpfe und Stadtentwicklung in Kopenhagen.* Berlin: Assoziation A.

Blomley, N. 2004. *Unsettling the City: Urban Land and the Politics of Property.* London and New York: Routledge.

Blomley, N. 2010. *Rights of Passage: Sidewalks and the Regulation of Public Flow.* London and New York: Routledge.

Bock, M. et al., eds. 1989. *Zwischen Resignation und Gewalt: Jugendprotest in den achtziger Jahren.* Opladen: Leske & Budrich.

Böcklemann, F. and Nagel, H. 1976. *Subversive Aktion: Die Sinn der Organisation ist ihr Scheitern.* Frankfurt/M: Neue Kritik.

Bodenschatz, H., Heise, V. and Korfmacher, J. 1983. *Schluss mit der Zerstörung? Stadterneuerung und städtische Opposition in Amsterdam, London und West-Berlin.* Giessen: Anabas.

Bojadžijev, M. 2008. *Die windige Internationale: Rassismus und Kämpfe der Migration.* Münster: Westfälisches Dampfboot.

Bojadžijev, M. 2015. "Housing, Financialisation, and Migration in the Current Global Crisis: An Ethnographically Informed View from Berlin." *SAQ: South Atlantic Quarterly*, 114: 29–45. DOI: 10.1215/00382876-2831268.

Bojadžijev, M. and Perinelli, M. 2010. "Die Herausforderung der Migration: Migrantische Lebenswelten in der Bundesrepublik in den siebziger Jahnren." In *Bundesrepublik Deutschland und Europa, 1968–1983*, edited by S. Reichardt and D. Siegfried. Göttingen: Wallstein Verlag, pp. 131–145.

Borneman, J. 1992. *Belonging in the Two Berlins: Kin, State, Nation.* Cambridge: Cambridge University Press.

Borneman, J. 1993. "Uniting the German Nation: Law, Narrative and Historicity." *American Ethnologist*, 20: 288–311. DOI: 10.1525/ae.1993.20.2.02a00050.

Bosse, P. and Zimmer, V., eds. 1988. *Ökologische Maßnahmen in Frauenstadtteilzentrums Schokoladenfabrik.* Berlin: Stern.

Böttcher, L. et al. 1978. "Strategie für Kreuzberg." *Arch +* 37: 63–73.

Bourdieu, P. 1977. *Outline of a Theory of Practice*. Cambridge: Cambridge University Press.

Boyle, M.S. 2011. "Berlin cracks down on squatters." http://www.counterpunch.org/2011/02/04/berlin-cracks-down-on-squatters/

Boyle, M.S. 2012. "The Ambivalence of Resistance: West German Antiauthoritarian Performance in an Age of Affluence." PhD Dissertation, University of California, Berkeley.

Brand, K.W., Busser, D. and Rucht, D. 1986. *Neue Soziale Bewegungen in West Europa und den USA*. Frankfurt/M: Campus Verlag.

Brand, R. and Fregonese, S. 2013. *The Radicals' City: Urban Environment, Polarisation, Cohesion*. London: Ashgate.

Brandes, V. and Schön, B., eds. 1981. *Wer sind die Instandbesetzer: Selbstzeugnisse, Dokumente, Analysen*. Bensheim: päd.-extra-Buchverlag.

Brennan, T. 2004. *The Transmission of Affect*. Ithaca, NY: Cornell University Press.

Brenner, N., Marcuse, P. and Mayer, M., eds. 2011. *Cities for People, Not for Profit: Critical Urban Theory and the Right to the City*. Oxford: Wiley-Blackwell.

Brenner, N. and Schmid, C. 2011. "Planetary Urbanization." In *Urban Constellations*, edited by M. Gandy. Berlin: Jovis Verlag, pp. 10–13.

Brenner, N. and Schmid, C. 2014. "The 'Urban Age' in Question." *International Journal of Urban and Regional Research*, 38: 731–755. DOI: 10.1111/1468-2427.12115.

Brickell, K. Forthcoming. "The World is Watching: Intimate Geopolitics of Forced Eviction and Women's Activism in Cambodia." *Annals of the Association of American Geographers*.

Brown, G. 2007. "Mutinous Eruptions: Autonomous Spaces of Radical Queer Activism." *Environment and Planning A*, 29: 2685–2698. DOI: 10.1068/a38385.

Brown, G. and Pickerill, J. 2009. "Space for Emotion in the Spaces of Activism." *Emotion, Space and Society* 2: 24–35. DOI: 10.1016/j.emospa.2009.03.004.

Brown, G. and Yaffe, H. 2013. "Practices of Solidarity: Opposing Apartheid in the Centre of London." *Antipode*, 46: 34–52. DOI: 10.1111/anti.12037.

Brown, T. 2009. "Music as a Weapon? Ton Steine Scherben and the Politics of Rock in Cold War Berlin." *German Studies Review*, 32: 1–22.

Brown, T. 2013. *West Germany and the Global Sixties: The Antiauthoritarian Revolt, 1962–1978*. Cambridge: Cambridge University Press.

Brückner, P. 2001 [1977]. *Ulrike Marie Meinhof und die deutschen Verhältnisse*. Berlin: Wagenbach.

Buck, H.F. 2004. *Mit hohem Anspruch gescheitert: Die Wohnungspolitik der DDR*. Münster: Lit Verlag.

Butler, J. 2002. "Afterword: After Loss, What Then?" In *Loss: The Politics of Mourning*, edited by D. Eng and D. Kazanjian. Berkeley, CA: University of California Press, pp. 467–473.

Butler, J. 2012. "So What are the Demands? And Where Do They Go From Here?" *Tidal: Journal of Occupy Theory* 2: 8–11.

Butler, J. and Spivak, G. 2007. *Who Sings the Nation-State: Language, Politics, Belonging*. Chicago, IL: University of Chicago Press.

Canning, K. 2002. *Languages of Labor and Gender: Female Factory Work in Germany, 1850–1914*. Ann Arbor, MN: The University of Michigan Press.

Canning, K. 2006. *Gender History in Practice: Historical Perspectives on Bodies, Class and Citizenship*. Ithaca, CA: Cornell University Press.

Carini, M. 2003. *Fritz Teufel: Wenn der Wahrheitsfindung dient*. Hamburg: Konkret Literatur Verlag.

Castells, M. 1983. *The City and the Grassroots*. Berkeley, CA: University of California Press.

Castillo, G. 2005. "Domesticating the Cold War: Household Consumption as Propaganda in Marshall Plan Germany." *Journal of Contemporary History*, 40: 261–288. DOI: 10.1177/0022009405051553.

Castillo, G. 2010. *Cold War on the Home Front: The Soft Power of Midcentury Design*. Minneapolis, MN: University of Minnesota Press.

Chamayou, G. 2012. *Manhunts: A Philosophical History*. Princeton, NJ: Princeton University Press.

Chatterton, P. and Hodkinson, S. 2006. "Autonomy in the City? Reflections on the Social Centres Movement in the UK." *City*, 10: 305–315. DOI: 10.1080/13604810600982222.

Chaussy, U. 1999. *Die drei Leben des Rudi Dutschke: Eine Biographie*. Zürich: Pendo Verlag.

Cochrane, A. 1999. "Re-imagining Berlin: World City, National Capital or Ordinary Place?" *European Urban and Regional Studies*, 6: 145–164. DOI: 10.1177/09697764900600204.

Colomb, C. 2011. *Staging the New Berlin. Place Marketing and the Politics of Urban Reinvention Post-1989*. London and New York: Routledge.

Colomb, C. 2012. "Pushing the Urban Frontier: Temporary Uses of Space, City Marketing, and the Creative City Discourse in 2000s Berlin." *Journal of Urban Affairs*, 34: 131–152. DOI: 10.1111/j.1467-9906.2012.00607.x.

Colvin, S. 2009. *Ulrike Meinhof and West German Terrorism: Language, Violence and Identity*. Rochester, NY: Camden House.

Cook, M. 2013. "'Gay Times': Identity, Locality, Memory, and the Brixton Squats in 1970s London." *Twentieth Century British History*, 24: 84–109. DOI: http://dx.doi.org/10.1093/tcbh/hwr053.

Cupers, K. and Miessen, M. 2002. *Spaces of Certainty*. Wuppertal: Müller + Busmann.

Datta, A. 2008. "Architecture of Low-income Widow Housing: 'Spatial Opportunities' in Madipur, West Delhi." *Cultural Geographies*, 15: 231–253. DOI: http://dx.doi.org/10.1177/1474474007087500.

Datta, A. 2012. *The Illegal City: Space, Law and Gender in a Delhi Squatter Settlement*. London: Ashgate.

Davies, A. 2012. "Assemblage and Social Movements: Tibet Support Groups and the Spatialities of Political Organisation." *Transactions of the Institute of British Geographers*, 37: 273–286. DOI: 10.1111/j.1475-5661.2011.00462.x.

Davis, B. 2000. *Home Fires Burning: Food, Politics, and Everyday Life in World War I Berlin*. Chapel Hill, NC: University of North Carolina Press.

Davis, B. 2006. "New Leftists and West Germany: Fascism, Violence, and the Public Sphere, 1967–1974." In *Coping with the Nazi Past: West German Debates on Nazism and Generational Conflict, 1955–1975*, edited by P. Gassert and A.E. Steinweis. New York: Berghahn, pp. 210–237.

Davis, B. 2008. "The City as Theater of Protest: West Berlin and West Germany." In *The Spaces of the Modern City: Imaginaries, Politics, and Everyday Life*, edited by G. Prakash and K.M. Krause. Princeton, NJ: Princeton University Press, pp. 247–274.

Davis, B. et al., eds. 2010. *Changing the World, Changing Oneself: Political Protest and Collective Identities in West Germany and the US in the 1960s and 1970s*. New York, NY: Berghahn Books.

Davis, M. 2006. *Planet of Slums*. London: Verso.

Delclós, C. 2013. "Victims No Longer: Spain's Anti-eviction Movement." https://www. opendemocracy.net/opensecurity/carlos-delcl%C3%B3s/victims-no-longer-spain% E2%80%99s-anti-eviction-movement

Deleuze, G. and Guattari, F. 1994. *What is Philosophy?* New York, NY: Columbia University Press.

Della Porta, D. 2006. *Social Movements, Political Violence, and the State.* Cambridge: Cambridge University Press.

Dellwo, K-H. and Baer, W. 2012. *Wir wollen alles: Häuserkampf I (1970–1989).* Hamburg: Laika Verlag.

Dellwo, K-H. and Baer, W. 2013. *Wir wollen alles: Häuserkampf II: Die Hausbesetzungen in Hamburg.* Hamburg: Laika Verlag.

Dennert, G. et al. 2007. *In Bewegung bleiben.* Berlin: Querverlag.

Dikeç, M. 2007. *Badlands of the Republic: Space, Politics, and Urban Policy.* Oxford: Wiley-Blackwell.

von Dirke, S. 1997. *'All Power to the Imagination': The West German Counterculture from the Student Movement to the Greens.* Lincoln: University of Nebraska Press.

Doron, G. 2000. "The Dead Zone and the Architecture of Transgression." *City* 4: 247–263. DOI: 10.1080/13604810050147857.

Dosse, F. 2011. *Gilles Deleuze and Félix Guattari: Intersecting Lives.* New York, NY: Columbia University Press.

Dressen, W., Kunzelmann, D. and Siepmann, E. 1991. *"Nilpferd des höllischen Urwalds. Spuren in eine unbekannte Stadt. Situationisten, Gruppe SPUR, Kommune I : ein Ausstellungsgeflecht des Werkbund-Archivs Berlin zwischen Kreuzberg und Scheunenviertel.* Berlin: Werkbund Archiv.

Durkheim, E. 1995 [1912]. *The Elementary Forms of Religious Life.* New York: Free Press.

Dutschke, R. 1968. "Von Antisemitismus zum Antikommunismus." In *Rebellion der Studenten oder Die Neue Opposition,* edited by U. Bergmann et al. Reinbek bei Hamburg, pp. 58–93.

Dutschke, R. 1981. *Aufrecht gehen: Eine fragmentarische Autobiographie.* Berlin: Olle and Wolter.

Dutschke-Klotz, G. 1996. *Rudi Dutschke – Wir hatten ein barbarisches, schönes Leben.* Cologne: Kiepenheuer & Witsch.

Eberstadt, R. 1920. *Handbuch des Wohnungswesens und der Wohnungsfrage.* Jena: Fischer.

Eley, G. 1989. "Labour History, Social History, *Alltagsgeschichte*: Experience, Culture and the Politics of Everyday Life." *Journal of Modern History,* 61: 297–343.

Eley, G. 2002. "Foreword." In *The Challenge of Modernity: German Social and Cultural Studies,* edited by A. von Saldern. Ann Arbor, MN: The University of Michigan Press, ix–xx.

Engels, F. 1995 [1872]. "The Housing Question." http://www.marxists.org/archive/marx/ works/1872/housing-question/index.htm

Enzensberger, U. 2004. *Die Jahre der Kommune I: Berlin, 1967–1969.* Cologne: Kiepenheuer & Witsch.

Ermittlungsausschuss im Mehringhof, ed. 1981a. *abgeräumt? 8 Häuser geräumt… Klaus-Jürgen Rattay tot. Eine Dokumentation.* Berlin: Mehringhof.

Ermittlungsausschuss im Mehringhof, ed. 1981b. *Dokumentation: Dezember Berlin 1980.* Berlin: Mehringhof.

Evangelische Kirche in Berlin-Kreuzberg. 1973. *2. Erklärung zur Stadterneuerung in Berlin-Kreuzberg*. Berlin: Kirchenkreis Koelln Stadt.

Fahlenbrach, K. 2002. *Protest-Inszeneirungen: Visuelle Kommunikation und kollektive Identitäten in Protestbewegungen*. Wiesbaden: Westdeutscher Verlag.

Featherstone, D. 2012. *Solidarity: Hidden Histories and Geographies of Internationalism*. London: Zed Books.

Featherstone, D. 2013. "Black Internationalism, Subaltern Cosmopolitanism, and the Spatial Politics of Antifascism." *Annals of the Association of America Geographers*, 103: 1406–1420. DOI: 10.1080/00045608.2013.779551.

Feigenbaum, A., Frenzel, F. and McCurdy, P. 2013. *Protest Camps*. London: Zed Books.

Felsmann, B. and Gröschner, A., eds. 2012. *Durchgangszimmer Prenzlauer Berg: Eine Berliner Künstlersozialgeschichte der 1970er und 1980er Jahre in Selbstauskünften*. Berlin: Lukas Verlag.

Fernández Arrigiotia, M. 2014. "Break-down: Undoing Home through Lifts and Stairs in a Puerto Rican Public Housing Demolition." *Home Cultures*, 11: 167–196. DOI: 10.275 2/175174214X13891916944634.

Fezer, J. and Heyden, M., eds. 2004. *Hier entsteht: Strategien partizipative Architektur und räumlicher Aneignung*. Berlin: b_books.

Fichter, T. and Lönnendonker, S. 1977. *Kleine Geschichte des SDS: Der Socialistische Deutsche Studentenbund von 1946 bis zur Selbstauflösung*. Berlin: Rotbuch.

Fields, D. and Uffer, S. Forthcoming. "The Financialisation of Rental Housing: A Comparative Analysis of New York City and Berlin." *Urban Studies*.

Ford, E. and Smith, J. 2014. "Häuserkämpfe: An Inside Look at Researching DIY Archives." http://www.inthelibrarywiththeleadpipe.org/2014/hauserkampfe-an-inside-look-at-researching-in-diy-archives/

Foucault, M. 2011. *The Courage of Truth: The Government of Self and Others, 1983–1984*. London: Palgrave Macmillan.

Foucault, M. 2000. "Lives of Infamous Men." In *Power: The Essential Works*, edited by J. Faubion. Volume 3. New York: The New Press, pp. 157–175.

Frisby, D. 2001. *Cityscapes of Modernity*. Cambridge: Polity Press.

Fritzsche, P. 1996. *Reading Berlin 1900*. Cambridge, MA: Harvard University Press.

Fulbrook, M. 2005. *The People's State: East German Society from Hitler to Honecker*. New Haven & London: Yale University Press.

Gay, P. 2001 [1968]. *Weimar Culture: The Outsider as Insider*. Harmondsworth: Penguin.

Garrett, B. 2013. *Explore Everything: Place-Hacking the City*. London: Verso.

Geist, J-F. and Kürvers, K. 1980–89. *Das Berliner Mietshaus*. Three volumes: 1980, 1984, 1989. Munich: Prestel.

Georg von Rauch-Haus. 1972. *Georg von Rauch-Haus: Kämpfen, Lernen, Leben*. Berlin: Kreuzberg Jugendzentrum.

Geronimo. 1992. *Feuer und Flamme: Zur Geschichte der Autonomen*. ID-Verlag: Berlin.

Geronimo. 2014. "Foreword." In *The City is Ours: Squatting and Autonomous Movements in Europe from the 1970s to the Present*, edited by B. van der Steen, A. Katzeff, and L. van Hoogehuijze. Oakland, CA: PM Press, pp. xii–xix.

Gilcher-Holtey, I., Kraus, D. and Scholer, I., eds. 2006. *Politisches-Theater nach 1968: Regie, Dramatik und Organisation*. Frankfurt/M: Campus Verlag.

Gilcher-Holtey, I., ed. 1998. *1968: Von Ereignis zum Gegenstand der Geschichtswissenschaft*. Göttigen: Vandenhoeck & Ruprecht.

Glaeser, A. 2000. *Divided in Unity: Identity, Germany and the Berlin Police*. Chicago, IL: University of Chicago Press.

Glaeser, A. 2011. *Political Epistemics: The Secret Police, the Opposition, and the End of East German Socialism*. Chicago, IL: University of Chicago Press.

Glaser, H. 2000. *Deutsche Kultur: Ein historischer Überblick von 1945 bis zur Gegenwart*. Bonn: Bundeszentrale für Politische Bildung.

Glatzer, R. 1993. *Berlin wird Kaiserstadt: Panorama einer Metropole, 1871–1890*. Berlin: Siedler Verlag.

Glomb, R. 1979. "Auf nach TUNIX. Collagierte Notizen zur Legitmationskrise des Staates." In *Gegenkultur heute: Die Alternativebewegung von Woodstock bis TUNIX*, edited by J. Gehre. 2nd ed. Amsterdam: Azid Presse, pp. 137–144.

Gothe, L. and Kippe, R. 1975. *Aufbruch, 5 Jahre Kampf des SSK: von der Projektgruppe für geflohene Fürsorgezöglinge über die Jugendhilfe zur Selbsthilfe verlendeter junger Arbeiter*. Cologne: Kiepenheuer & Witsch.

Gould, D. 2009. *Moving Politics: Emotion and ACT UP's Fight against Aids*. Chicago, IL: University of Chicago Press.

Graham, S. 2010. *Cities under Siege: The New Military Urbanism*. London: Verso.

Grashoff, U. 2011a. *Leben in Abriss: Schwarzwohnen in Halle an der Saale*. Halle: Hasenverlag.

Grashoff, U. 2011b. *Schwarzwohnen: Die Unterwanderung der staatlichen Wohnraumlenkung in der DDR*. Göttingen: V&R unipress.

Gregory, D. 2006. "Introduction: Troubling Geographies." In *David Harvey: A Critical Reader*, edited by N. Castree and D. Gregory. Oxford: Wiley-Blackwell, pp. 1–25.

Groth, J. and Corijn, E. 2005. "Reclaiming Urbanity: Indeterminate Spaces, Informal Actors and Urban Agenda Setting." *Urban Studies*, 42: 503–526. DOI: 10.1080/00420980500035436.

Guattari, F. 1986. *Les années d'hiver, 1980–1985*. Paris: Barrault.

Guattari, F. 2009 [1978]. "New Spaces of Liberty for Minoritarian Desire." In *Soft Subversions: Texts and Interviews, 1977–1985*, F. Guattari. Los Angeles, CA: Semiotext(e), pp. 94–101.

Guattari, F. and Lotringer, S. 2009 [1982]. "A New Alliance is Possible." In *Soft Subversions: Texts and Interviews, 1977–1985*, F. Guattari. Los Angeles, CA: Semiotext(e), pp. 113–127.

Guattari, F. and Rolnik, S. 2008. *Molecular Revolution in Brazil*. Los Angeles, CA: Semiotext(e).

Gutmair, U. 2013. *Die ersten Tagen von Berlin: Der Sound der Wende*. Stuttgart: Klett-Cotta.

Häberlen, J. and Smith, J. 2014. "Struggling for Feelings: The Politics of Emotions in the Radical New Left in West Germany." *Contemporary European History*, 23: 615–637. DOI: 10.1017/S0960777314000289.

Habermas, J. 1988. "Der Marsch durch die Insttutionen hat auch die CDU erreicht." *Frankfurt Rundschau*, 11.3.1988.

Hake, S. 2008. *Topographies of Class: Modern Architecture and Mass Society in Weimar Berlin*. Ann Arbor, MN: The University of Michigan Press.

Halbrock, C. 2004. "Vom Widerstand zum Umbruch: die oppositionelle Szene in den 80er Jahren." In *Prenzlauer Berg im Wandel der Geschichte: Leben rund um den Helmholtzplatz*, edited by B. Roder and B. Tacke. Berlin: be.bra verlag, pp. 98–124.

Haller, M. 1981. "Aussteigen oder rebellieren. Über die Doppeldeutigkeit der Jugendrevolte." In *Aussteigen oder rebellieren: Jugendliche gegen Staat und Gesellschaft*, edited by M. Haller. Hamburg: Rowohlt, pp. 7–22.

Hannah, M. 2010. *Dark Territory in an Information Age: Learning from the West German Census Controversies of the 1980s*. London: Ashgate.

Hannah, M. 2012. "Foucault's 'German Moment': Genealogy of a Disjuncture." *Foucault Studies*, 13: 116–137.

Hansen, M. 1995. "America, Paris, the Alps: Kracauer (and Benjamin) on Cinema and Modernity." In *Early Cinema and the Invention of Modern Life*, edited by L. Charney and V. Schwartz. Berkeley, CA: University of California Press, pp. 363–403.

Hanshew, K. 2012. *Terror and Democracy in West Germany*. Cambridge: Cambridge University Press.

Hardt, M. 1996. "Laboratory Italy." In *Radical Thought in Italy: A Potential Politics*, edited by M. Hardt and P. Virno. Minneapolis, MN: University of Minnesota Press, pp. 1–10.

Hardt, M. and Negri, A. 2009. *Commonwealth*. Cambridge, MA: Harvard University Press.

Härlin, B. 1981. "Von Haus zu Haus – Berliner Bewegungsstudien." *Kursbuch*, 65: 1–28.

Harvey, D. 1982. *The Limits to Capital*. London: Verso.

Harvey, D. 1985. *The Urbanisation of Capital*. Oxford: Blackwell.

Harvey, D. 1989. *The Urban Experience*. Baltimore, MD: The Johns Hopkins University Press.

Harvey, D. 2008. "The Right to the City." *New Left Review*, 53: 23–40.

Harvey, D. 2012. *From the Right to the City to the Urban Revolution*. London: Verso.

Harvey, D. 2014. *Seventeen Contradictions and the End of Capitalism*. London: Profile Books.

Häusermann, H. and Siebel, W. 1996. *Soziologie des Wohnens: Eine Einführung in Wandel und Ausdifferenzierung des Wohnens*. Weinheim & München: Juventa Verlag.

Haydn, F. and Temel, R., eds. 2006. *Temporäre Räume: Konzepte zur Stadtnutzung*. Basel: Birkhauser.

Hegemann, W. 1979 [1930]. *Das steinerne Berlin: Geschichte der größten Mietskasernenstadt der Welt*. Braunschweig: Vieweg.

Hentschel, C. Forthcoming. "Postcolonialising Berlin and the Fabrication of the Urban." *International Journal of Urban and Regional Research*.

Herzog, D. 2005. *Sex after Fascism: Memory and Morality in Twentieth-Century Germany*. Princeton, NJ: Princeton University Press.

Hermann, J. 1925. *Geschichte der deutschen Mieterbewegung*. Dresden: Bund Dt. Mieter.

Heyden, M. 2008. "Evolving Participatory Design: A Report from Berlin, Reaching Beyond." *Field Journal*, 2: 31–46.

Heyden, M. n.d. "Die Freiraumgestaltung der K 77." Unpublished manuscript.

Heyden, M. and Schaber, I. 2008. "Here is the Rose, Here is the Dance!" In *Who Says Concrete Doesn't Burn, Have You Tried? West Berlin Film in the '80s*, edited by S. Strasthaus and F. Wüst. Berlin: b_books, pp. 132–148.

Hinze, A.M. 2013. *Turkish Berlin: Integration Policy and Urban Space*. Minneapolis, MN: University of Minnesota Press.

von Hodenberg, C. and Siegfried, D. 2006. "Mass Media and the Generation of Conflict: West Germany's Long Sixties and the Formation of a Critical Public Sphere." *Contemporary European History*, 15: 367–395. DOI: 10.1017/S0960777306003377.

Hoffmann-Axthelm, D. et al., eds. 1979. *Zwei Kulturen: TUNIX, Mescalero und die Folgen*. Berlin: Verlag Ästhetik und Kommunikation.

Höhn, M. 2002. *GIs and Fräuleins: The German-American Encounter in 1950s West Germany*. Chapel Hill, NC: University of North Carolina Press.

Höhn, M. 2008. "The Black Panther Solidarity Committees and the Voice of the Lumpen." *German Studies Review*, 31: 133–154.

Holm, A. 2006. *Die Restrukturierung des Raumes: Stadterneuerung der 90er Jahre in Ostberlin.* Bielefeld: transcript Verlag.

Holm, A. 2010. *Wir bleiben alle! Gentrifizierung-Städtische Konflikte um Aufwertung und Verdrängung.* Münster: Unrast Verlag.

Holm, A. 2014a. *Mietenwahnsinn: Warum wohnen immer teurer wird und wer davon profitiert.* Munich: Knaur.

Holm, A. 2014b. "Reclaim Berlin." In *Reclaim Berlin: Soziale Kämpfe in der neoliberalen Stadt*, edited by A. Holm. Berlin: Assoziation A, pp. 7–24.

Holm, A. and Kuhn, A. 2010. "Häuserkampf und Stadterneuerung." *Blättern für deutsche und internationale Politik*, March: 107–115.

Holm, A. and Kuhn, A. 2011. "Squatting and Urban Renewal: The Interaction of Squatter Movements and Strategies of Urban Sestructuring in Berlin." *International Journal of Urban and Regional Research*, 35: 644–658. DOI: 10.1111/j.1468-2427.2010.001009.x.

Holston, J. 2008. *Insurgent Citizenship: Disjunctions of Democracy and Modernity in Brazil.* Princeton, NJ: Princeton University Press.

Huber, J. 1980. *Wer soll das alles ändern? Die Alternativen und die Alternativbewegung.* Berlin: Rotbuch.

Huyssen, A. 1997. "The Voids of Berlin." *Critical Inquiry*, 24: 57–81.

Ingold, T. 2000. *Perceptions of the Environment, Essays on Livelihood, Dwelling, and Skill.* London and New York: Routledge.

Iveson, K. 2007. *Publics and the City.* Oxford: Wiley-Blackwell.

Iveson, K. 2013. "Cities Within the City: Do-it-yourself Urbanism and the Right to the City." *International Journal of Urban and Regional Research*, 37: 941–956. DOI: 10.1111/1468-2427.12053.

Jackson, S. 2000. *Lines of Activity: Performance, Historiography, Hull-House Domesticity.* Ann Arbor: The University of Michigan Press.

Jacobs, J. and Merriman, P. 2011. "Practicing architectures." *Social and Cultural Geography*, 12: 211–222. DOI: 10.1080/14649365.2011.565884.

Jelavich, P. 1993. *Berlin Cabaret.* Cambridge, MA: Harvard University Press.

Juchler, I. 1996. *Die Studentenbewegungen in den Vereinigten Staaten und der Bundesrepublik Deutschland der sechziger Jahre.* Berlin: Duncker & Humblot.

Juris, J. 2008. *Networking Futures: The Movements against Corporate Globalisation.* Durham, NC: Duke University Press.

Kadir, N. 2014. "Myth and Reality in the Amsterdam Squatters' Movement." In *The City is Ours: Squatting and Autonomous Movements in Europe from the 1970s to the Present*, edited by B. van der Steen, A. Katzeff, and L. van Hoogehuijze. Oakland, CA: PM Press, pp. 21–61.

Kaes, A. et al., eds. 1994. *The Weimar Republic Sourcebook.* Berkeley, CA: University of California Press.

Kanngieser, A. 2007. "Gestures of Everyday Resistance: The Significance of Play and Desire in the Umsonst Politics of Collective Appropriation." http://eipcp.net/transversal/0307/kanngieser/en

Kanngieser, A. 2012. "…And…And…And…The Transversal Politics of Performative Encounters." *Deleuze Studies*, 6: 265–290. DOI: 10.3366/dls.2012.0062.

Kanngieser, A. 2013. *Experimental Politics and the Making of Worlds.* London: Ashgate.

Karakayali, S. 2000. "Across Bockenheimer Landstraße." *Diskus 2/00*, www.copyriot.com/diskus/2_00/a.htm

Karakayali, S. 2005. "Lotta Continua in Frankfurt, Türken Terror in Köln. Migrantische Kämpfe in der Geschichte der Bundesrepublik." *Grundrisse*, http://www.grundrisse.net/grundrisse14/14serhat_karakayali.htm

Karakayali, S. 2009. *Gespenster der Migration: zur Genealogie illegaler Einwanderung in der Bundesrepublik Deutschland*. Bielefeld: transcript.

Karapin, R. 2007. *Protest Politics in Germany: Movements on the Left and Right since the 1960s*. University Park, PA: PSU Press.

Katsiaficas, G. 2006. *The Subversion of Politics: European Autonomous Social Movements and the Decolonisation of Everyday Life*. Oakland, CA: AK Press.

Kätzel, U. 2002. *Die 68erinnen: Porträt einer rebellischen Frauengeneration*. Hamburg: Rowohlt.

Katz, S. and Mayer, M. 1985. "Gimme Shelter: Self-help Housing Struggles Within and Against the State in New York City and Berlin." *International Journal of Urban and Regional Research*, 9: 15–46. DOI: 10.1111/j.1468-2427.1985.tb00419.x.

Kerngehäuse Cuvrystraße. 1980. *Gewerbehof Cuvrystraße 20/23: Leben und arbeiten in SO 36*. Berlin: Initiative 'Kerngehäuse Cuvrystraße'.

Kersting, F-W. 2006. "Juvenile Left-wing Radicalism, Fringe Groups, and Anti-psychiatry in West Germany." In *Between Marx and Coca-Cola*, edited by A. Schildt and D. Siegfried. New York and Oxford: Berghahn, pp. 353–375.

Kirchknopf, G. 1968. "Vom elastischen Familienverband zur Kommune." *Kursbuch*, 14: 110–115.

Klein, J. and Porn, S. 1981. "Instandbesetzen." In *Besetzung – weil das Wünschen nicht geholfen hat*, edited by I. Müller-Münch et al. Hamburg: Rowohlt, pp. 108–125.

Klemek, C. 2011. *The Transatlantic Collapse of Urban Renewal: Postwar Urbanism from New York to Berlin*. Chicago, IL: University of Chicago Press.

Klimke, M. and Scharloth, K., eds. 2007. *1968: Handbuch zur Kultur – und Mediengeschichte der Studentenbewegung*. Stuttgart: J.B. Metzler.

Klimke, M. 2009. "'We Are Not Going to Defend Ourselves Before Such a Justice System!' – 1968 and the Courts." *German Law Journal*, 10: 261–274.

Klimke, M. 2010. *The "Other Alliance": Global Protest and Student Unrest in West Germany and the US, 1962–1972*. Princeton, NJ: Princeton University Press.

von Klöden, K.F. 1874. *Jugenderinerungen*. Berlin: DT. Verlag.

Kluge, A. 1973. *Lernprozesse mit tödlichem Ausgang*. Frankfurt/M: Suhrkamp Verlag.

Kluge, A. and Negt, O. 2014 [1981]. *History & Obstinacy*. Cambridge, MA: Zone Books.

Knorr, P. 1969. "Bei Kommunarden in Berlin und Hamburg." *Pardon*, 5: 26–40.

Koenen, G. 2001. *Das rote Jahrzehnt: Unsere kleine deutsche Kulturrevolution 1967–1977*. Frankfurt/M: Fischer Verlag.

Kolling, S. 2008. *Honig aus dem zweiten Stock: Berliner Hausprojekte erzählen*. Berlin: Assoziation A.

Kommune 2. 1969. *Versuch der Revolutionierung des bürgerlichen Individuums*, Berlin: Oberbaumverlag.

König, T. 1965. "'Von der Gewalt,' Partial translation of Frantz Fanon's *The Wretched of the Earth*." *Kursbuch*, 2: 1–55.

Kontányi, A. and Vaneigem, R. 1961. "Basic Program of the Bureau of Unitary Urbanism." http://www.bopsecrets.org/SI/6.unitaryurb.htm

Koopmans, R. 1995. *Democracy from Below: New Social Movements and the Political System in West Germany*. Oxford: Westview Press.

Korczak, D. 1981. *Rückkehr in die Gemeinschaft: Kleine Netze: Berichte über Wohnsiedlungen*. Frankfurt/M: Fischer.

Koshar, R. 2000. *From Monuments to Traces: Artifacts of German Memory, 1870–1990*. Berkeley, CA: University of California Press.

Kotti & Co. 2014. "Kotti & Co und das Recht auf Stadt." In *Reclaim Berlin: Soziale Kämpfe in der neoliberalen Stadt*, edited by A. Holm. Berlin: Assoziation A, pp. 343–354.

Kowalczuk, I-S. 1992. "Wohnen ist wichtiger als das Gesetz." Historische Streflichter zu Wohnungsnot und Mieterwiderstand in Berlin." In *Berliner, Mainzer Straße*, edited by S. Arndt et al. Berlin: Basisdruck, pp. 231–259.

Kowalczuk, I-S. 2002. *Freiheit und Öffentlichkeit: Politischer Samisdat in der DDR 1985 bis 1989*. Berlin: Robert-Havemann-Gesellschaft.

Kowalczuk, I-S. et al., 2006. *"Für ein freies Land mit freien Menschen." Opposition und Widerstand in Biographien und Fotos*. Berlin: Robert-Havemann-Gesellschaft.

Krahl, H-J. 1971. *Konstitution und Klassenkampf: Zur historischen Dialektik von bürgerlicher Emanzipation und proletarischer Revolution*. Frankfurt/M: Verlag Neue Kritik.

Kraus, D. 2007. *Theater-Protest: Zur Politisierung von Straße und Bühne in den 1960er Jahren*. Frankfurt: Campus Verlag.

Kraushaar, W. 1978. *Autonomie oder Getto? Kontroversen über die Alternativebewegung*. Frankfurt: Verlag Neue Kritik.

Kraushaar, W. 1998. *Frankfurter Schule und Studentenbewegung - Von der Flaschenpost zum Molotowcocktail*. Bd. I-III. Hamburg: Rogner & Bernhard.

Kraushaar, W. 2000. *1968 als Mythos, Chiffre und Zäsur*. Hamburg: Hamburg Edition.

Kuhn, A. 2005. *Stalins Enkel, Maos Söhne: die Lebenswelt der K-Gruppen in der Bundesrepublik der 70er Jahre*. Frankfurt/M: Campus.

Kuhn, G. 2012. *All Power to the Councils! A Documentary History of the German Revolution*. Oakland, CA: PM Press.

Kunst & KulturCentrum Kreuzberg. 1984. *Dokumentation*. Berlin: KuKuCK.

Kunzelmann, D. 1998. *Leisten sie keinen Widerstand! Bilder aus meinem Leben*. Berlin: Transit Schwarzenbach.

Ladd, B. 1990. *Urban Planning and Civic Order in Germany, 1860–1914*. Cambridge, MA: Harvard University Press.

Ladd, B. 1997. *The Ghosts of Berlin: Confronting German History in the Urban Landscape*. Chicago, IL: University of Chicago Press.

Lang, B. 1998. *Mythos Kreuzberg: Ethnographie eines Stadtteils (1961–1995)*. Frankfurt/M: Campus Verlag.

Lange, A. 1976. *Berlin zur Zeit Bebels und Bismarcks: Zwischen Reichsgründung und Jahrhundertwende*. Berlin: Dietz.

Langhans, R. 2008. *Ich bin's: Die ersten 68 Jahre. Autobiographie*. Berlin: Blumenbar.

Langhans, R. and Teufel, F. 1968. *Klau Mich*. Berlin: Edition Verlag.

Laqueur, W. 1974. *Weimar: A Cultural History, 1918–1933*. New York, NY: Putnam.

Larsson, B. 1967. *Demonstrationen: Ein Berliner Modell. Enstehung der demokratischen Opposition*. Berlin: Voltaire.

Laurisch, B. 1981. *Kein Abriß unter dieser Nummer*. Berlin: Anabas Verlag.

Lazzarato, M. 2009. *Expérimentations politiques*. Paris: Éditions Amsterdam.

Lee, M. 2007. "Umherschweifen und Spektakel: Die situationistische Tradition." In *1968: Ein Handbuch zur Kultur- und Mediengeschichte der Studentenbewegung*, edited by M. Klimke and J. Scharloth. Stuttgart: Metzler-Verlag, pp. 101–106.

Lee, M. 2011. "Gruppe Spur: Art as a Revolutionary Medium during the Cold War." In *Between the Avant Garde and the Everyday: Subversive Politics in Europe, 1958–2008*, edited by T. Brown and L. Anton. New York: Berg Books, pp. 11–30.

Lees, A. 1985. *Cities Perceived: Urban Society in European and American Thought, 1820–1940*. New York, NY: Columbia University Press.

Lees, L. 2001. "Towards a Critical Geography of Architecture: The Case of an Ersatz Colosseum." *Ecumene*, 8: 51–86. DOI: 10.1177/096746080100800103.

Lefebvre, H. 1991. *The Production of Space*. Oxford: Blackwell.

Lefebvre, H. 1996. "Right to the City." In *Writing on Cities*, edited by E. Kofman and E. Lebas. Oxford: Blackwell, pp. 61–184.

Lefebvre, H. 2009. *State, Space, World: Selected Essays*. Minneapolis, MN: University of Minnesota Press.

Lefebvre, H. 2014 [1976]. *Toward an Architecture of Enjoyment*. Minneapolis, MN: University of Minnesota Press.

Lindenberger, T. 1984. "Berliner Unordnung zwischen den Revolutionen." In *Pöbelexzesse und Volkstumulte in Berlin: Zur Sozialgeschichte der Straße (1830–1980)*, edited by M. Gailus. Berlin: europäische perspektiven, pp. 43–78.

Linker, G. 1981. "Hausbesetzungen – ein Rückblick auf die Zukunft?" *Polizei, Technik, Verkehr*, 1, pp. 85ff.

Lönnendonker, S. et al., eds. 2002. *Die antiautoritäre Revolte: Der Sozialistische Deutsche Studentenbund nach der Trennung von der SPD*. Wiesbaden: Westdeustscher Verlag.

Lönnendonker, S. and Fichter, T., eds. 1975. *Freie Universität Berlin, 1948–1973: Hochschule im Umbruch*, vol. 4. Berlin: Pressestelle der Freien Universität.

López, M.M. 2013. "The Squatters' Movement in Europe: A Durable Struggle for Social Autonomy in Urban Politics." *Antipode*, 45: 866–887. DOI: 10.1111/j.1467-8330. 2012.01060.x.

Lüdtke, A. 1993. *Eigen-Sinn: Fabrikalltag, Arbeiterfahrungen und Politik vom Kaiserreich bis in den Faschismus*. Hamburg: Ergebnisse.

Lüdtke, A., ed. 1995. *The History of Everyday Life: Reconstructing Historical Experiences and Ways of Life*. Princeton, NJ: Princeton University Press.

MacDougall, C. 2011a. "Cold War Capital: Contested Urbanity in West Berlin, 1963–1989." PhD Dissertation, Rutgers University.

MacDougall, C. 2011b. "In the Shadow of the Wall: Urban Space and Everyday Life in Kreuzberg." In *Between the Avant-Garde and the Everyday: Subversive Politics in Europe from 1957 to the Present*, edited by T. Brown and L. Anton. New York: Berg, pp. 154–173.

Mandel, R. 2008. *Cosmopolitan Anxieties: Turkish Challenges to Citizenship and Belonging in Germany*. Durham, NC: Duke University Press.

Marcuse, H. 1955. *Eros and Civilization*. Boston: Beacon Press.

Marcuse, H. 1964. *One-Dimensional Man*. Boston: Beacon Press.

Marcuse, H. 1965. *Repressive Tolerance*. Boston: Beacon Press.

Markovits, A. and Gorski, P. 1993. *The German Left: Red, Green and Beyond*. Oxford: Oxford University Press.

Marwick, A. 1998. *The Sixties Cultural Revolution in Britain France Italy and the United States, c.1958–c.1974*. Oxford: Oxford University Press.

März, M. 2012. *Linker Protest nach dem Deutschen Herbst: Eine Geschichte des linken Spektrums im Schatten des 'starken Staates', 1977–1979*. Bielefeld: transcript.

Massumi, B. 2002. *Parables for the Virtual: Movement, Affect, Sensation*. Durham, NC: Duke University Press.

Mayer, M. 2009. "The 'Right to the City' in the Context of Shifting Mottos of Urban Social Movements." *City*, 13: 362–374. DOI: 10.1080/13604810902982755.

Mayer, M. 2013a. "First World Urban Activism." *City*, 17: 5–19. DOI:10.1080/13604813. 2013.757417.

Mayer, M. 2013b. "Preface." In *Squatting in Europe: Radical Spaces, Urban Struggles*, edited by SqEK. London: Minor Compositions, pp. 1–9.

McCormick, R. 1991. *Politics of the Self: Feminism and the Postmodern in West German Literature and Film*. Princeton, NJ: Princeton University Press.

McElligott, A., ed. 2001. *The German Urban Experience, 1900–1945*. London and New York: Routledge.

McFarlane, C. 2008. "Sanitation in Mumbai's Informal Settlements: State, 'Slum' and Infrastructure." *Environment and Planning A*, 26: 480–499. DOI: 10.1068/a39221.

McFarlane, C. 2011a. "The City as Assemblage: Dwelling and Urban Space." *Environment and Planning D: Society and Space*, 29: 649–671. DOI: 10.1068/d4710.

McFarlane, C. 2011b. *Learning the City: Knowledge and Translocal Assemblage*. Oxford: Wiley-Blackwell.

McFarlane, C. and Vasudevan, A. 2013. "Informal Infrastructures." In *Handbook of Mobilities*, edited by P. Adey et al. London and New York: Routledge, pp. 256–264.

Mega Spree 2010. "Wer wir sind. Was wir wollen." http://www.megaspree.de

Meinhof, U. 1971. "Bambule: Fürsorge – Sorge für wen?." *Rotbuch*, 24.

Meinhof, U. 2008. *Everybody Talks About the Weather… We Don't: The Writings of Ulrike Meinhof*. New York, NY: Seven Stories Press.

Melucci, A. 1980. "The New Social Movements: A Theoretical Approach." *Social Science Information*, 19: 199–226. DOI: 10.1177/053901848001900201.

Melucci, A. 1988. "Getting Involved: Identity and Mobilisation in Social Movements." *International Social Movement Research*, 1: 329–348.

Melucci, A. 1996. *Challenging Codes: Collective Action in the Information Age*. Cambridge: Cambridge University Press.

Merrifield, A. 2013a. "Citizens' Agora: The New Urban Question." *Radical Philosophy*, May/June, 179: 31–35.

Merrifield, A. 2013b. *The Politics of the Encounter: Urban Theory and Protest under Planterary Urbanization*. Athens, GA: University of Georgia Press.

Merrifield, A. 2013c. "The Urban Question under Planetary Urbanization." *International Journal of Urban and Regional Research*, 37: 909–922. DOI: 10.1111/j.1468-2427. 2012.01189.x.

Merrifield, A. 2014. *The New Urban Question*. London: Pluto Press.

Mikkelsen, F. and Karpantschof, R. 2001. "Youth as a Political Movement: Development of the Squatters' and Autonomous Movement in Copenhagen, 1981–1995." *International Journal of Urban and Regional Research*, 25: 609–628. DOI: 10.1111/1468-2427.00332.

Miller, B. and Nicholls, W. 2013. "Social Movements in Urban Society: The City as a Space of Politicization." *Urban Geography*, 34: 452–473. DOI: 10.1080/02723638.2013.786904.

Misselwitz, P. et al. 2007. "Stadtentwicklung ohne Städtebau: Planerischer Alptraum oder gelobtes Land?" In *Urban Pioneers. Berlin: Stadtentwicklung durch Zwischennutzung*.

Temporary Use and Urban Development in Berlin, edited by SenStadt Senatsverwaltung für Stadtentwicklung. Berlin: Architektenkammer & Jovis Verlag, pp. 102–109.

Mitchell, D. 2003. *The Right to the City: Social Justice and the Fight for Public Space*. New York: Guilford Press.

Mitchell, P. n.d. "Socialism's Contested Urban Space: A Study of East German Squatters." Unpublished manuscript.

Mitchell, W.J.T. 2012. "Image, Space, Revolution: The Arts of Occupation." *Critical Inquiry*, 39: 8–32. DOI: 10.1086/668048.

Mitscherlich, A. 1965. *Die Unwirtlichkeit unserer Städte*. Frankfurt/M: Suhrkamp Verlag.

Möbius, P. 1973. "Kinderkultur." *Kursbuch*, 34: 25–48.

Moeller, R., ed. 1997. *West Germany Under Construction: Politics, Society, and Culture in the Adenauer Era*. Ann Arbor: The University of Michigan Press.

Moldt, D., ed. 2005. *Der mOaning star 1985–1989: eine Ostberliner Untergrundpublikation*. Berlin: Robert-Havemann-Gesellschaft.

Moldt, D. 2008. *Die Blues-Messen: Ein Kapital des Widerstandes in der DDR*. Berlin: Robert-Havemann-Gesellschaft.

Molé, N. 2012. *Labour Disorders in Neoliberal Italy: Mobbing, Well-Being and the Workplace*. Bloomington, IN: University of Indiana Press.

Mudu, P. 2004. "Resisting and Challenging Neoliberalism: The Development of Italian Social Centres." *Antipode*, 36: 917–941. DOI: 10.1111/j.1467-8330.2004.00461.x.

Muñoz, J.E. 2009. *Cruising Utopia: The Then and There of Queer Futurity*. New York, NY: NYU Press.

Negri, A. 2002. "Multitude and Metropolis." http://www.generation-online.org/t/metropolis.htm

Negri, A. 2005. *Books for Burning: Between Civil War and Democracy in 1970s Italy*. London: Verso.

Negri, A. 2013. *The Winter is Over: Writings on Transformation Denied, 1989–1995*. Los Angeles, CA: Semiotext(e).

Negt, O. 1995. *Achtundsechzig: Politische Intellektuele und die Macht*, Göttingen: Steidl.

Neuwirth, R. 2006. *Shadow Cities: A Billion Squatters, a New Urban World*. London and New York: Routledge.

Nicholls, W.J. 2008. "The Urban Question Revisited: The Importance of Cities for Social Movements." *International Journal of Urban and Regional Research*, 32: 841–859. DOI: 10.1111/j.1468-2427.2008.00820.x

Niethammer, L. and Bruggermeier, F-J. 1976. "Wie wohnten Arbeiter im Kaiserreich?." *Archiv für Sozialgeschichte*, 16: 61–134.

Nitsche, R. et al., eds. 1981. *Häuserkämpfe, 1872, 1920, 1945, 1982*. Berlin: Campus Verlag.

Notz, G. 2006. *Warum flog die Tomate? Die autonomen Frauenbewegungen der Siebzigerjahr*. Ulm: AG Spak.

Novak, P., ed. 2014. *Zwangsräumungen verhindern: Ob nuriye ob kalle, Wir blieben alle*. Berlin: Assoziation A.

Novy, J. and Colomb, C. 2013. "Struggling for the Right to the (Creative) City in Berlin and Hamburg: New Urban Social Movements, New 'Spaces of Hope'?" *International Journal of Urban and Regional Research*, 37: 1816–1838. DOI: 10.1111/j.1468-2427.2012.01115.x.

Oswalt, P. and Stegers, R. 2000. *Berlin: Stadt ohne Form*. Munich: Prestel.

Ottersky, L. 1957. "Zum Problem der Sanierung 'was wird aus unseren überalterten Stadtteilen?'" *Bauwelt*, 19: 457.

Owens, L. 2009. *Cracking under Pressure: Narrating the Decline of the Amsterdam Squatters' Movement*. Amsterdam: Amsterdam University Press.

Papenbrock, M. 2007. "Happening, Fluxus, Performance." In *1968: Handbuch zur Kultur – und Mediengeschichte der Studentenbewegung*, edited by M. Klimke and J. Scharloth. Stuttgart: J.B. Metzler, pp. 137–149.

Péchu, C. 2010. *Les Squats*. Paris: Les Presses des Sciences Po.

Peck, J. 2012a. "Recreative City: Amsterdam, Vehicular Ideas and the *Adaptive Spaces of Creativity Policy*." *International Journal of Urban and Regional Research*, 36: 462–485. DOI: 10.1111/j.1468-2427.2011.01071.x.

Peck, J. 2012b. "Austerity Urbanism." *City*, 16: 626–655. DOI: 10.1080/13604813.2012.734071

Pfaff, S. 2006. *Exit-Voice Dynamics and the Collapse of East Germany: The Crisis of Leninism and the Revolution of 1989*. Durham, NC: Duke University Press.

Pieterse, E. 2008. *City Futures: Confronting the Crisis of Urban Development*. London: Zed Books.

Pinder, D. 2005. *Visions of the City: Utopianism, Power and Politics in Twentieth-Century Urbanism*. Edinburgh: University of Edinburgh Press.

Pithouse, R. 2006. "Thinking Resistance in the Shanty Town." http://www.metamute.org/editorial/articles/thinking-resistance-shanty-town

Plowman, A. 1998. *The Radical Subject: Social Change and the Self in Recent German Autobiography*. Bern: Peter Lang.

Poling, K. 2014. "Shantytowns and Pioneers Beyond the City Wall: Berlin's Urban Frontier in the Nineteenth Century." *Central European History*, 47: 245–274. DOI: 10.1017/S0008938914001241.

Pruijt, H. 2003. "Is the Institutionalization of Urban Movements Inevitable? A Comparison of the Opportunities for Sustained Squatting in New York and Amsterdam." *International Journal of Urban and Regional Research*, 27: 133–157. DOI: 10.1111/1468-2427.00436.

Pruijt, H. 2013. "The Logic of Urban Squatting." *International Journal of Urban Regional Research*, 37: 19–45. DOI: 10.1111/j.1468-2427.2012.01116.x.

Pugh, E. 2014. *Architecture, Politics and Identity in Divided Berlin*. Pittsburgh, PA: University of Pittsburgh Press.

Purcell, M. 2003. "Citizenship and the Right to the Global City: Reimagining the Capitalist World Order." *International Journal of Urban and Regional Research*, 27: 564–590. DOI: 10.1111/1468-2427.00467.

Rabehl, B. 2002. "Die Provokationselite: Aufbruch und Scheitern der subversive Rebellion in den Sechziger Jahren." In *Die antiautoritäre Revolte: Der Sozialistische Deutsche Studentenbund nach der Trennung von der SPD*, edited by S. Lönnendonker et al. Wiesbaden: Westdeustscher Verlag, pp. 400–512.

Rada, U. 1991. *Mietenreport: Alltag, Skandale und Widerstand*. Berlin: Ch. Links Verlag.

Rada, U. 1997. *Berlin: Hauptstadt der Verdrängung: Berliner Zukunft zwischen Kiez und Metropole*. Berlin: Assoziation A.

Rampf, C. 2009. *Hausfriedensbruch: §123 StGB*. Berlin: Berliner-Wissenschafts-Verlag.

Rancière, J. 2012a. *Proletarian Nights: The Workers' Dream in Nineteenth-Century France*, new ed. London: Verso.

Rancière, J. 2012b. "Talk on Proletarian Nights." New York, 18 September 2012.

Raunig, G. 2007. *Art and Revolution: Transversal Activism in the Long Twentieth Century*. Los Angeles, CA: Semiotext(e).

Raunig, G. 2010. *A Thousand Machines*. Los Angeles, CA: Semiotext(e).

Reddy, W. 2001. *The Navigation of Feeling: A Framework for the History of Emotions*. Cambridge: Cambridge University Press.

Regenbogenfabrik Block 109 e.V. 2006. *Festschrift zum 25. Jubiläum der Regenbogenfabrik*. Berlin: Kreuzberg Museum.

Reichardt, S. and Siegfried, D., eds. 2010. *Das Alternative Milieu: Antibürgerliche Lebenstil und linke Politik in der Bundesrepublik Deutschland und Europa, 1968–1983*. Göttingen: Wallstein Verlag.

Reichardt, S. 2014. *Authentizität und Gemeinschaft: Linksalternatives Leben in den siebziger und frühen achtzier Jahren*. Frankfurt/M: Suhrkamp Verlag.

Reimann, A. 2009. *Dieter Kunzelmann: Avantgardist, Protestler, Radikaler*. Göttingen: Vandenhoeck and Ruprecht.

Rethmann, P. 2011. "West German Radical History Seen from the 1980s: *Projekt Artur* and the Refusal of Political Confiscation." *Seminar: A Journal of Germanic Studies*, 47: 46–63. DOI: 10.3138/seminar.47.1.46

Reynolds, B. 2009. *Transversal Subjects: From Montaigne to Deleuze after Derrida*. Basingstoke, UK: Palgrave Macmillan.

Ribbe, W., ed. 2002. *Geschichte Berlins*, 2 Vols. Berlin: Berlin Wiss.-Verlag.

Ring, M. 1872. "Ein Besuch in Barackia: Berliner Lebensbild." *Gartenlaube*, 72: 458–461.

Ritchie, A. 1999. *Faust's Metropolis: A History of Berlin*. New York, NY: Harper Collins.

Romanos, E. 2014. "Evictions, Petitions and Escraches: Contentious Housing in Austerity Spain." *Social Movement Studies*, 13: 296–302. DOI: 10.1080/14742837.2013.830567.

Rosenbladt, S. 1981. "Die "Legalos" von Kreuzberg." in *Hausbesetzer: Wofür sie kämpfen, wie sie leben und wie sie leben wollen*, edited by S. Aust and S. Rosenbladt. Hamburg: Hoffmann und Campe, pp. 28–51.

Ross, K. 1988. *The Emergence of Social Space: Rimbaud and the Paris Commune*. Minneapolis, MN: University of Minnesota Press.

Ross, K. 2002. *May '68 and its Afterlives*. Chicago, IL: University of Chicago Press.

Rothe, K. 2013. *Betongold: Wie die Finanzkrise in mein Wohnzimmer kam* (Video, 53 min).

Roy, A. 2011. "Slumdog Cities: Rethinking Subaltern Urbanism." *International Journal of Urban and Regional Research* 35: 223–238. DOI: 10.1111/j.1468-2427.2011.01051.x.

Rucht, D. 1996. "German Unification, Democratization and the Role of Social Movements: A Missed Opportunity." *Mobilization*, 1: 35–62.

Rühle, O. 1930. *Illustrierte Kultur und Sittengeschichte des Proletariats*. Berlin: Neue Deutscher Verlag.

Ryder, A. 1967. *The German Revolution of 1918: A Study of German Socialism in War and Revolt*. Cambridge: Cambridge University Press.

von Saldern, A. 1984. *Auf dem Wege zum Arbeiter-Reformismus: Parteialltag in sozialde-mokratischer Provinz: Göttingen, 1870–1920*. Frankfurt/M: Materalis.

von Saldern, A. 1993. *Neues Wohnen: Wohnungspolitik und Wohnkultur im Hannover der Zwanziger Jahre*. Hannover: Hahnsche Buchhandlung.

von Saldern, A. 1995. *Häuserleben: Zur Geschichte städtischen Arbeiterwohnens vom Kaiserreich bis heute*. Bonn: J.H.W. Dietz.

Sassen, S. 2014. *Expulsions: Brutality and Complexity in the Global Economy*. Cambridge, MA: Harvard University Press.

Sevilla-Buitrago, A. 2011. "'This Square is our Home!' The Organization of Urban Space in the Spanish 15-M Movement." *Progressive Planning*, 189: 42–49.

Scarpa, L. 1986. *Martin Wagner und Berlin: Architektur und Städtebau in der Weimarer Republik*. Wiesbaden: Vieweg Verlag.

Scharenberg, A. and Bader, I. 2009. "Berlin's Waterfront Site Struggle." *City*, 13, 325–335. DOI: 10.1080/13604810902982938.

Scharloth, J. 2010. *1968: Eine Kommunikationsgeschichte*. Munich: Wilhelm Fink.

Scheer, J. and Espert, J. 1982. *'Deuschland, Deutschland, alles ist vorbei:' Alternatives Leben oder Anarchie? Die neue Jugendrevolte am Beispiel der Berliner 'Scene'*. Munich: Berhard and Graefe.

Scheffler, K. 1989 [1910]. *Berlin – ein Stadtschicksal*. Berlin: Fannel & Walz.

Scherer, K-J. 1984. "Berlin (West): Hauptstadt der szenen: Ein Portrait kultureller und anderer Revolten Anfang der achtziger Jahre." In *Pöbelexzesse und Volkstumulte in Berlin: Zur Sozialgeschihte der Strasse 1830–1980*, edited by M. Gailus. Berlin: Verlag Europäische Perspektiven, pp. 197–222.

Schling, H. 2013. "Eviction Brixton: Creating Housing Insecurity in London." https://www.opendemocracy.net/opensecurity/hannah-schling/eviction-brixton-creating-housing-insecurity-in-london

Schneider, P. 1969. "Die Phantasie im Spätkapitalismus und die Kulturrevolution." *Kursbuch*, 16: 1–37.

Schneider, P. 2010. *Rebellion und Wahn: Mein '68*. Cologne: Kiepenheuer&Witsch.

Schön, R. 1982. "Besetzung leerstehende Häuser – Hausfriedensbruch?" *Neue Juristische Wochenschrift*, 35: 1126–1129.

Schultze, T. and Gross, A. 1997. *Die Autonomen: Ursprünge, Entwicklung und Profil der Autonome Bewegung*. Hamburg: Konkret Verlag.

Schwedler, R. 1964. "Stadterneuerung in Berlin. Senator für Bau-und Wohnungswesen." Meeting, Abgeordnetenhaus, Berlin. 18 June.

Seibert, N. 2008. *Vergessene Proteste: Internationalismus und Antirassismus, 1964–1983*. Münster: Unrast Verlag.

Seidel, W., ed. 2006. *Scherben: Musik, Politik und Wirkung der Ton Stein Scherben*. Mainz: Ventil Verlag.

Senator für Bau- und Wohnungswesen. 1957. *Auf dem halben Wege ... von der Mietskaserne zum sozialen Wohnungsbau*. Berlin: Druckhaus Tempelhof.

Sennett, R. 1970. *The Uses of Disorder: Personal Identity and City Life*. Harmondsworth: Pelican Books.

Shaw, K. 2005. "The Place of Alternative Culture and the Politics of its Protection in Berlin, Amsterdam and Melbourne." *Planning Theory & Practice*, 6: 149–169. DOI: 10.1080/14649350500136830.

Sheridan, D. 2007. "The Space of Subculture in the City: Getting Specific about Berlin's Indeterminate Territories." *Field Journal*, 1: 97–119.

Shukaitis, S. 2007. *Imaginal Machines: Autonomy & Self-Organisation in the Revolutions of Everyday Life*. London: Minor Compositions.

Sichtermann, K., Johler, J. and Stahl, C. 2000. *Keine Macht für Niemand: Die Geschichte der 'Ton Steine Scherben'*. Berlin: Schwarzkopf & Schwarzkopf.

Siedler, J. 1965. *Die gemordete Stadt: Abgesang auf Putte und Strasse, Platz und Baum*. Berlin: Herbig.

Sieg, K. 2002. *Ethnic Drag: Performing Race, Nation, Sexuality in West Germany*. Ann Arbor, MN: The University of Michican Press.

Siegfried, D. 2006. "Protest am Markt. Gegenkultur in der Konsumgesellschaft um 1968." In *Wo 1968 liegt: Reform und Revolte in der Geschichte der Bundesrepublik*, edited by C. von Hodenberg and D. Siegfried. Göttingen: Vendenhoeck and Ruprecht, pp. 48–78.

Siegfried, D. 2008. *Sound der Revolte: Studien zur Kulturrevolution um 1968.* Weinheim: Juventa.

Silver, J. 2014. "Incremental Infrastructures: Material Improvisation and Social Collaboration across Post-colonial Accra." *Urban Geography*, 35: 788–804. DOI: 10.1080/02723638.2014.933605.

Simone, A.M. 2004. *For the City Yet to Come: Changing African Life in Four Cities.* Durham, NC: Duke University Press.

Simone, A.M. 2005. "Urban Circulation and the Everyday Politics of African Urban Youth: The Case of Douala Cameroon." *International Journal of Urban and Regional Research*, 29: 516–532. DOI: 10.1111/j.1468-2427.2005.00603.x.

Simone, A.M. 2008. "Emergency Democracy and the 'Governing' Composite." *Social Text*, 26: 13–33. DOI: 10.1215/01642472-2007-027.

Simone, A.M. 2010. *City Life from Jakarta to Dakar: Movements at the Crossroads.* London and New York: Routledge.

Simone, A.M. 2014. *Jakarta: Drawing the City Near.* Minneapolis, MN: University of Minnesota Press.

Slobodian, Q. 2012. *Foreign Front: Third World Politics in Sixties West Germany.* Durham, NC: Duke University Press.

Slobodian, Q. 2013a. "The Borders of the Rechtsstaat in the Arab Autumn: Deportation and Law in West Germany, 1972/1973." *German History*, 31: 204–224. DOI: 10.1093/gerhis/ght019.

Slobodian, Q. 2013b. "Bandung in Divided Germany: Managing Non-aligned Politics in East and West, 1955–1963." *The Journal of Commonwealth History*, 41: 644–662. DOI: 10.1080/03086534.2013.840438.

Smith, N. 1996. *The New Urban Frontier: Gentrification and the Revanchist City.* London and New York: Routledge.

Sozialistische Selbsthilfe Köln. 1981. *Sanierung macht Angst, Angst macht krank, Sanierung macht krank: Eine Dokumentation.* Cologne: SSK.

Sonnewald, B. and Raabe-Zimmerman, J. 1983. *Die 'Berliner Linie' und die Hausbesetzer-Szene.* Berlin: Berlin Verlag.

Soukup, U. 2007. *Wie starb Benno Ohnesorg: Der 2. Juni 1967.* Berlin: Verlag 1900.

SqEK – Squatting Europe Kollektive, eds. 2013. *Squatting in Europe: Radical Spaces, Urban Struggles.* London: Minor Compositions.

SqEK – Squatting Europe Kollektive, eds. 2014. *The Squatters' Movement in Europe: Commons and Autonomy as Alternative to Capitalism.* London: Pluto Books.

Staggenborg, S. 2001. "Beyond Culture Versus Politics: A Case Study of a Local Women's Movement." *Gender & Society*, 15: 507–530. DOI: 10.1177/089124301015004002.

Stephens, R.P. 2007. *Germans on Drugs: The Complications of Modernization in Hamburg.* Ann Arbor, MN: The University of Michigan Press.

Strategien für Kreuzberg. 1978. *Strategien für Kreuzberg* Berlin: Der Senator für Bau- u. Wohnungswesen.

Strohmeyer, K. 2000. *James Hobrecht und die Modernisierung der Stadt.* Berlin: Wissenschafts-Verlag.

Süss, W. and Rylewski, R. 1999. *Berlin: Die Hauptstadt.* Berlin: Nicolai.

Suttner, A. 2011. *'Beton brennt': Hausbesetzer und Selbstverwaltung im Berlin, Wien und Zürich der 80er.* Münster: LIT Verlag.

Tannerfeldt, G. and Ljung, P. 2006. *More Urban, Less Poor: An Introduction to Urban Development and Management.* London: Earthscan.

Tarrow, S. 1989. *Democracy and Disorder.* Oxford: Oxford University Press.

Tarrow, S. 1998. *Power in Movement: Social Movements and Contentious Politics.* Cambridge: Cambridge University Press.

Thomas, N. 2003. *Protest Movements in 1960s West Germany: A Social History of Dissent and Democracy.* Oxford and New York: Berg.

Till, K. 2005. *The New Berlin: Memory, Politics, Place.* Minneapolis, MN: University of Minnesota Press.

Till, K. 2011. "Interim Use at a Former Death Strip? Art, Politics and Urbanism at Skulpturenpark Berlin_Zentrum." In *After the Wall: Berlin in Germany and Europe*, edited by M. Silberman. Basingstoke: Palgrave Macmillan, pp. 99–122.

Tilly, C. 1995. *Popular Contention in Great Britain, 1758–1834.* Cambridge, MA: Harvard University Press.

Tilly, C. 2008. *Contentious Performance.* Cambridge: Cambridge University Press.

Tilly, C. and Tarrow, S. 2006. *Contentious Politics.* Oxford: Oxford University Press.

Tisdall, C. 1974. *Art into Society, Society into Art.* London: ICA.

Tomann, F. 1996. "Germany." In *Housing Policy in Europe*, edited by P. Balchin. London and New York: Routledge, pp. 51–66.

Tompkins, A. n.d. "Particularly Universal? Black Panther Party, Angela Davis, and the West German Left." Unpublished manuscript.

Touraine, A. 1981. *The Voice and the Eye: An Analysis of Social Movements.* Cambridge: Cambridge University Press.

Tronti, M. 1965. "The Strategy of Refusal." http://libcom.org/library/strategy-refusalmario-tronti

Turner, J. 1976. *Housing by People: Towards Autonomy in Building Environments.* London: Marion Boyars.

Turner, V. 1969. *The Ritual Process: Structure and Anti-Structure.* Chicago: Aldine Publishing.

Uffer, S. 2014. "Wohnungsprivatisierung in Berlin: Eine Analyse verscheidener Investitionsstrategien und deren Konsequenze für die Stadt und ihre Bewohner." In *Reclaim Berlin*, edited by A. Holm. Berlin: Assoziation A, pp. 64–82.

Uitermark, J. 2004. "The Co-optation of Squatters in Amsterdam and the Emergence of a Movement Meritocracy: A Critical Reply to Pruijt." *International Journal of Urban and Regional Research*, 28: 687–698. DOI: 10.1111/j.0309-1317.2004.00543.x.

Uitermark, J. 2011. "An Actually Existing Just City? The Fight for the Right to the City in Amsterdam." In *Cities for People, Not for Profit: Theory/Practice*, edited by N. Brenner, P. Marcuse and M. Meyer. Oxford: Wiley-Blackwell, pp. 197–214.

Urban, F. 2013. "The Hut on the Garden Plot: Informal Architecture in Twentieth-Century Berlin." *Journal of the Society of Architectural Historians*, 72: 221–249. DOI: 10.1525/jsah.2013.72.2.221.

von Uslar, M. 1997. "Einer, der gern saß." *Der Spiegel*, 24: 72–81.

Van der Steen, B., Katzeff, A. and Van Hoogenhuijze, L., eds. 2014. *The City is Ours: Squatting and the Autonomous Movements in Europe from the 1970s to the Present.* Oakland, CA: PM Press.

Van Houdt, F. et al. 2011. "Neoliberal Communitarian Citizenship: Current Trends Towards 'Earned Citizenship' in the United Kingdom, France and the Netherlands." *International Sociology*, 26: 408–432. DOI: 10.1177/0268580910393041.

Van Schipstal, I.L.M. and Nicholls, W. 2014. "Rights to the Neoliberal City: the Case of Urban Land Squatting in 'Creative' Berlin." *Territory, Politics, Governance*, 2: 173–193. DOI: 10.1080/21622671.2014.902324

Varon, J. 2004. *Bringing the War Home: The Weather Underground, the Red Army Faction, and Revolutionary Violence in the Sixties and Seventies*. Berkeley, CA: University of California Press.

Vasudevan, A. 2005. "Metropolitan Theatrics: Performing the Modern in Weimar Berlin." PhD Dissertation, University of British Columbia.

Vasudevan, A. 2006. "Experimental Urbanisms: Psychotechnik in Weimar Berlin." *Environment and Planning D: Society and Space*, 24: 799–826. DOI: 10.1068/d375t.

Vasudevan, A. 2007. "Symptomatic Acts, Experimental Embodiments: Theatres of Scientific Protest in Weimar Germany." *Environment and Planning A*, 1812–1837. DOI: 10.1068/a38295.

Vasudevan, A. 2011a. "Dramaturgies of Dissent: The Spatial Politics of Squatting in Berlin, 1968–." *Social and Cultural Geography*, 12: 283–303. DOI: 10.1080/14649365.2011.564734.

Vasudevan, A. 2011b. "Legal Activism: The Spatial Politics of Squatting in the UK." https://libcom.org/library/legal-activism-spatial-politics-squatting-uk

Vasudevan, A. 2013. "Schwarzwohnen: The Spatial Politics of Squatting in East Berlin." https://www.opendemocracy.net/opensecurity/alexvasudevan/schwarzwohnen-spatial-politics-of-squatting-in-east-berlin

Vasudevan, A. 2014a. "The Autonomous City: Towards a Critical Geography of Occupation." *Progress in Human Geography*. DOI:10.1177/0309132514531470.

Vasudevan, A. 2014b. "The Makeshift City: Towards a Global Geography of Squatting." *Progress in Human Geography*. DOI:10.1177/0309132514531471.

Voss, K. 1996. "The Collapse of a Social Movement: The Interplay of Mobilizing Structures, Framing, and Political Opportunities in the Knights of Labor." In *Comparative Perspectives on Social Movements: Political Opportunities, Mobilizing Structures, and Cultural Framings*, edited by D. McAdam, J. McCarthy and M. Zaid. Cambridge: Cambridge University Press, pp. 226–258.

Wacquant, L. 2007. "Territorial Stigmatization in the Age of Advanced Marginality." thesis eleven, 91: 66–77. DOI: 10.1177/0725513607082003.

Waits, N. and Wolmar, C., eds. 1980. *Squatting: The Real Story*. London: Bay Leaf Books.

Ward, C. 2002. *Cotters and Squatters: Housing's Hidden History*. Nottingham: Five Leaves.

Ward, J. 2011. *Post-Wall Berlin: Borders, Space and Identity*. Basingstoke: Palgrave Macmillan.

Weinhauer, K. et al. 2006. *Terrorismus in der Bundesrepublik: Medien, Staat und Subkulturen in den 1970er Jahren*. Frankfurt/M: Campus Verlag.

Weipert, A. 2013. *Das Rote Berlin: Eine Geschichte der Berliner Arbeiterbewegung, 1830–1934*. Berlin: BWV.

Weszkalnys, G. 2010. *Berlin Alexanderplatz: Transforming Place in a Unified Germany*. New York: Berghahn Books.

Wright, S. 2002. *Storming Heaven: Class Composition and Struggle in Italia Autonomist Marxism*. London: Pluto Press.

Yiftachel, O. 2009. "Theoretical Notes on 'Gray Cities': The Coming of Urban Apartheid." *Planning Theory*, 8: 88–100. DOI: 10.1177/1473095208099300.

Index

Metropolitan Preoccupations: The Spatial Politics of Squatting in Berlin, First Edition.
Alexander Vasudevan.
© 2015 John Wiley & Sons, Ltd. Published 2015 by John Wiley & Sons, Ltd.